T0259681

Textile Science and Clothing Technology

Series editor

Subramanian Senthilkannan Muthu, Kowloon, Hong Kong

More information about this series at http://www.springer.com/series/13111

Subramanian Senthilkannan Muthu
Editor

Sustainable Innovations in Textile Chemical Processes

 Springer

Editor
Subramanian Senthilkannan Muthu
Kowloon
Hong Kong

ISSN 2197-9863   ISSN 2197-9871   (electronic)
Textile Science and Clothing Technology
ISBN 978-981-13-4167-0   ISBN 978-981-10-8491-1   (eBook)
https://doi.org/10.1007/978-981-10-8491-1

© Springer Nature Singapore Pte Ltd. 2018
Softcover re-print of the Hardcover 1st edition 2018
This work is subject to copyright. All rights are reserved by the Publisher, whether the whole or part
of the material is concerned, specifically the rights of translation, reprinting, reuse of illustrations,
recitation, broadcasting, reproduction on microfilms or in any other physical way, and transmission
or information storage and retrieval, electronic adaptation, computer software, or by similar or dissimilar
methodology now known or hereafter developed.
The use of general descriptive names, registered names, trademarks, service marks, etc. in this
publication does not imply, even in the absence of a specific statement, that such names are exempt from
the relevant protective laws and regulations and therefore free for general use.
The publisher, the authors and the editors are safe to assume that the advice and information in this
book are believed to be true and accurate at the date of publication. Neither the publisher nor the
authors or the editors give a warranty, express or implied, with respect to the material contained herein or
for any errors or omissions that may have been made. The publisher remains neutral with regard to
jurisdictional claims in published maps and institutional affiliations.

Printed on acid-free paper

This Springer imprint is published by the registered company Springer Nature
Singapore Pte Ltd. part of Springer Nature
The registered company address is: 152 Beach Road, #21-01/04 Gateway East,
Singapore 189721, Singapore

This book is dedicated to:
The lotus feet of my beloved
Lord Pazhaniandavar
My beloved late Father
My beloved Mother
My beloved Wife Karpagam
and Daughters—Anu and Karthika
My beloved Brother
Last but not least
To everyone working in the global textile
chemical processing sector to make it
SUSTAINABLE

Contents

Chapter 1
Sustainable Dyeing Techniques

P. Senthil Kumar and P. R. Yaashikaa

Abstract The success of textile industry depends on the colour of the finished product which attracts humans. Textile industry often deals with design and conversion of yarn into fibre then into fabric finally dyed and fabricated into finished clothing. Dyeing is defined as the method of imparting colour to finished products mainly fabrics or during the initial stage to the yarn itself. Three main processes are involved in dyeing process namely preparation, dyeing and finishing. Dyeing process may be carried out in batch or in continuous mode. It consists of a special solution containing dye and chemical for binding. The two factors influencing dyeing process are temperature and time controlling. There are different methods applied by textile industries for adding colour to the products. Few dyeing techniques include exhaust, pad, pad-fixation, printing, bale, batik, beam, chain, cross, random, etc. Though dyes are attractive in nature, their impact on environment depends on the type of substance used and removal and degradation of dye substances. The contaminated wastewater discharged out after dyeing process contains huge amount of chemical substances which has negative impact on environment.

Keywords Textile industry · Dyeing process · Dyeing techniques
Degradation · Dyeing machines

1 Introduction

The textile industry is primarily examined about the plan and creation of yarn, fabric, apparel, and their dispatch. The crude material might be regular, or engineered utilizing the outcomes of other industries usually chemical. Textile industry is a standout amongst the most water-concentrated sectors. The effluent from textile

P. Senthil Kumar (✉) · P. R. Yaashikaa
Department of Chemical Engineering, SSN College of Engineering, Chennai 603110, India
e-mail: senthilchem8582@gmail.com

P. R. Yaashikaa
e-mail: yaashikaapr@ssn.edu.in

© Springer Nature Singapore Pte Ltd. 2018
S. S. Muthu (ed.), *Sustainable Innovations in Textile Chemical Processes*,
Textile Science and Clothing Technology, https://doi.org/10.1007/978-981-10-8491-1_1

industry is portrayed by high chemical oxygen demand (COD), biological oxygen demand (BOD), total dissolved solids (TDS) and pH. While yarn is for the most part created in the factories, textures are delivered in the power loom and handloom segments also. Ecological issues related with the textile industry are commonly those related with water contamination. Characteristic contaminations removed from the fibre being prepared alongside the chemicals utilized for handling are the two fundamental wellsprings of contamination [1]. Effluents released from textile industry are hot, basic, strong odour and coloured by chemicals which are used as a part of colouring process. A portion of the chemicals released are lethal. The main process of textile industry relays on the change of fibre into yarn, yarn into texture which are then coloured or printed and created into garments. Distinctive sorts of fibre are utilized to create yarn. Cotton remains the most imperative normal fibre, so is dealt with top to bottom. There are numerous variable procedures accessible at the turning and texture framing stages combined with the complexities of the completing and colouration procedures to the creation of wide scopes of outcomes. There remains a large scale industry that utilizes hand systems to accomplish similar outcomes [2].

The textile industry can be classified into two major divisions based on their development as composed and sloppy textile industries. Composed textile industries is an exceptionally sorted out one with enormous significance on the capital concentrated generation process. This segment is described by refined plants where innovatively propelled apparatus are used for large scale manufacturing of textile materials. Sloppy Textile Industry segment is the predominant piece of this industry which for the most part uses the conventional practices woven or spun in fabric creation and consequently is work concentrated in nature. This industry is described by the creation of garments either through weaving or turning with the assistance of hands. The decentralized nature is considered as another critical element of the sloppy textile industry in India [3].

Globalization is unpreventable and necessary under the present world financial circumstance. Numerous ventures are influenced decidedly or adversely by the globalization drift. There is no exception for the textile industry from this case. It has been confronting an emergency circumstance in a previous couple of years. It confronts an extreme rivalry on the global advertise. Indian textile industry is one of the biggest and the most vital areas in the economy as far as yield, outside trade profit, and business eras. India has likewise been a huge player in the worldwide textile markets. It is the third biggest maker of cotton, second biggest maker of silk, the biggest maker of jute and the fifth biggest maker of man-made strands and yarn. The Indian Textile Industry, being one of the most seasoned and critical divisions gains a great deal of remote trade and utilizes an impressive level of the community from both rural and urban ranges. Cotton exports have developed as a noteworthy wellspring of outside trade winning for the country. Indian Textile industry has been seeing a major basic change, persistently re-evaluating and rediscovering it to address the issues of the worldwide consumers. Indian Companies have begun raising their benchmarks and furthermore forcefully seeking their human asset methodologies [4]. The overall dyeing process of textile industry is shown in Fig. 1.

Fig. 1 Outline of dyeing
process

2   Dye and Its Importance

Dyes might be characterized as substances that, when connected to a substrate give
shading by a procedure that adjusts, in any event briefly, any gem structure of the
hued substances. Such substances with impressive dyeing limit are broadly utilized in
the textile industry, pharmaceutical, nourishment, cosmetics, plastics, photographic,
etc. The colours can hold fast to perfect surfaces by arrangement, by framing cova-
lent bond or forming complexes with metallic salts, through the process of physical
adsorption or by mechanical maintenance. The textile industry expands a generous
measure of water in its assembling forms utilized for the most part in the colouring
and completing or finishing operations. The wastewater from textile industrial plants
is named the most dirtying of all the mechanical segments, considering the quantity
created and in addition, the composition of effluents released. Likewise, the expanded

interest for textile outcomes and the corresponding increment in their generation, and the utilization of manufactured colours have together added to dye wastewater getting to be plainly one of the generous wellsprings of extreme contamination issues in at times. A standout amongst the most troublesome assignments defined by the wastewater treatment plants of the textile industry is the evacuation of the shade of these mixes, principally in light of the fact that colours and shades are intended to oppose biodegradation, with the end goal that they stay on the earth for a drawn out stretch of time. The colouring procedure is one of the key factors in the effective exchanging of textile outcomes [5]. The buyer typically searches for some essential item attributes, for example, great obsession regarding light, sweat, and washing, both at first and after delayed utilize. To guarantee these properties, the substances that offer to shade to the fibre must demonstrate a high resemblance, even shading, imperviousness to vanishing and be financially possible. Present day dyeing innovation comprises of a few stages chose by the idea of the fibre and properties of the colours and shades for use in textures, for example, chemical structure, arrangement, product accessibility, financial contemplations, etc. Colouring techniques have not changed much with time. Essentially, water is utilized to clean, colour and spread other chemicals to the textures, and furthermore to wash the treated fibres or textures. The fibres utilized as a part of the textile industry can be classified into two fundamental gatherings designated natural and synthetic fibres. Common Natural fibres are gotten from the earth plants or creatures, for example, fleece, cotton, flax, silk, jute, and sisal, the majority of which depend on cellulose and proteins. Then again, engineered or synthetic fibres are natural or organic polymers, generally got from oil sources, for instance, polyester, polyamide, rayon, etc. The two most significant materials in the textile industry are cotton, the biggest, and polyester [6].

3   Principles of Dyeing Process

The goal of dyeing is to deliver even colour of a substrate as a rule to coordinate pre-chosen shading. The shading ought to be uniform all through the substrate and be a strong shade with no unevenness or change in the shade over the entire substrate. There are many variables that will impact the presence of the last shade, including the surface of the substrate, development of the substrate by means of physical and chemical methods, pre-treatment connected to the substrate before coloring and post final treatments related after the coloring procedure. Material coloring includes the utilization of various distinctive chemicals and other agents to help the coloring procedure. Some of them are process-particular, while others are additionally utilized as a part of different operations. A few agents are as of now contained in the dyestuff plan; yet more ordinarily dispersing agents are added at a later stage to the dye solution. The fixing agents are usually found in the effluents as they do not retain to the substrate after the dyeing process is completed [7].

Fig. 2 Dyeing process steps

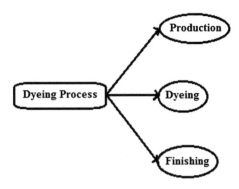

4   Steps Involved in Dyeing Process

The following are three steps involved in the dyeing process (Fig. 2).

4.1   Production

Production is the process in which undesirable contaminations are expelled from the textures previous to colouring. This can be done by cleaning with fluid basic substances and cleansers or by applying compounds like enzymes. Numerous textures are decolorized with hydrogen peroxide or chlorine-containing mixes to expel their regular shading, and if the texture is to be white in colour and not coloured, optical lighting up medium are included.

4.2   Dyeing

The colouring is the fluid use of adding colour to the textile substrates, basically utilizing manufactured or engineered colours and as often as possible at high temperatures and weights. It is essential to bring up that there is no colour which colours every current fibre and no fibre which can be coloured with every single known colour. Amid this progression, the colours and chemical substance help, for example, surfactants, acids, soluble base/bases, electrolytes, chelating operators, emulsifying oils, etc. are connected to the material to get a uniform profundity of shading with the shading quickness properties. This procedure incorporates dissemination of the colour into the fluid stage took after by adsorption onto the external surface of the fibres, lastly dispersion and adsorption on the inward surface of the strands. Based upon the normal end utilization of the textures, diverse quickness properties might be required. Distinctive sorts of dye and chemical added substances are utilized to get

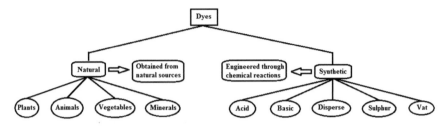

Fig. 3 Classification of dyes

these properties, which is completed amid the completing stage. Dyeing can likewise be refined by applying shades together with binding agents.

4.3 Finishing

The final completing step includes analysis with synthetic mixes of chemical substances for enhancing the nature of the texture. Changeless press analysis, water sealing, softening, antistatic assurance, soil resistance, recolors discharge and microbial/parasitic prevention is few methods of fabric analysis involved during the last finishing stage.

5 Classification of Dyes

The texture, properties and application of dyes are determined by their chemical structure forming the basis for their classification. A more normal classification of dyes and their related compounds involve the adequate binding to chemical structural compounds so that more similar compounds are placed close to one another. Basically dyes can be classified as "natural" obtained from vegetables, minerals, plant and animals and "chemical or synthetic" dyes synthesised using chemical compounds (Fig. 3).

5.1 Natural Dyes

Natural dyes deliver exceptionally unprecedented, mitigating and delicate shades when contrasted with engineered or synthetic dyes. The term 'natural dye' includes extraction of dye from all natural resources including plants, animals, minerals, etc. Natural dyes are generally non-substantive and must be connected to materials by the assistance of mordants, typically a metallic salt, having a liking for both the shading matter and the fiber. Metal ions usually possess more interrelating forces capable of binding with weak or moderate attraction forces at the time of interaction between the textile material and metallic salt. These metallic mordant subsequent to joining with color in the fiber, it shapes an insoluble hasten or lake and in this manner both the color and stringent get settled to end up wash quick to a sensible level. Natural colors are known for their utilization in the shading of nourishment substrate, leather and additionally characteristic fibres like fleece, silk and cotton as real zones of use since pre-memorable circumstances. In spite of the fact that this old craft of coloring materials with regular natural colors withstood the waste of time, because of the wide accessibility of engineered colors at a sparing value, a fast decrease in common coloring proceeded [8].

Advantages of Natural dyes

- Natural dyes are soft, renewable, biodegradable and safe to environment.
- It does not fade away with time.
- Utilization of chemical based synthetic dyes can be reduced.
- Labour intensive industry providing employment opportunities to many people in the process of extraction, cultivation and application of natural dyes.
- There is no problem of waste disposal from this industry since those wastes can be used as a biofertilizer in agricultural fields.

Drawbacks of Natural dyes

- Expensive and Time consuming process involves the requirement of knowledge and skilled labours in dyeing techniques.
- In case of extraction of dye from plants, synthesise of same shades is difficult process since crops may vary in season, time and place.
- All natural dyes require mordants for binding with the textile materials.

5.1.1 Types of Natural Dyes

Natural dyes can be classified according to the source they are derived from like vegetables, animals, plants and minerals, according to the mode of application as direct, acidic or basic dyes, according to their chemical structure and colours as yellow, red, black, orange, etc. (Table 1).

Table 1 Sources of natural dyes

S. no.	Source	Name
1	Plants	Catechu, Indigofera plant, Kamala tree, Madder root, etc.
2	Animals	Insects like Laccifer lacca, Cochineal, Murex snail, Octopus, etc.
3	Vegetables	Beetroot, Berries, etc.
4	Minerals	Ochre

5.2 Synthetic Dyes

Synthetic or Chemical dyes are manufactured by industries through chemical reactions such as oxidation, reduction or by using chromatographic methods. These dyes sometimes cause environmental pollution and health problems. Though all dyes fade to an extent, synthetic dyes fade quickly compared to natural dyes [9]. Few synthetic dyes are as follows:

5.2.1 Acid Dye

The term 'acid dye' starts from the coloring procedure, which is done in acidic (pH 2.0–6.0) fluid arrangement. Acid dyes or anionic dyes incorporate many mixes from the most changed chromophoric frameworks which show trademark contrasts in structure, however, have as a typical component water-solubilising ionic substituents. The anionic colors on a basic level incorporate direct colors, be that as it may, in view of their trademark structures these are utilized to color cellulose containing materials that are connected to the fibre from a dye bath. The gathering of anionic colors additionally incorporates a substantial extent of receptive colours, which notwithstanding the typical basic attributes likewise contain bunches that can respond with utilitarian gatherings of the fibre amid the coloring procedure. In order to increase their solubility nature, acid dyes possess aromatic molecules along with sulfonyl and amino group that are complex in structure. In textile industry, acid dyes are more suitable for silk, wool, acrylics, etc. and are comparatively less suitable for cellulosic fibres since they have low affinity towards cotton cellulose. Acid dyes bind with fibre strands by means of hydrogen bonds, Van der Waals force or ionic interactions [10].

5.2.2 Basic Dye

Basic dyes are usually positively charged cationic colouring molecules because of the presence of amino groups in their structure. Since they are positively charged, basic dyes reacts with negatively charged compounds. So binding occurs through ionic interactions only. Ionic holding occurs between negatively charged substrate and positively charged basic dyes. Consequently, basic dyes can be risky for recreational shading or for use in the home condition. Consequently, basic dyes can be risky for recreational shading or for use in the home condition. If not utilized with wellbeing and security alert, fundamental colors can recolor undesired materials like skin and body, any glass or plastic containers, etc. The chemical structure of basic dyes finds vast application in dyeing plastics, acrylics, textile industries, etc. Basic dyes are known for their extensive variety of shades, their colour and dynamic quality, and their similarity with manufactured, anionic materials. Basic dyes are exceedingly favoured when shading engineered, cationic materials, for example, acrylics. At the point when utilized with materials that are chemically better, basic dyes yield lively, brilliant, and enduring shades that other different sorts of dyes are not ready to accomplish [11].

5.2.3 Disperse Dye

Disperse dyes are the main water-insoluble colors that color polyester and acetic acid derivation strands. Disperse dye particles are the smallest among all dye molecules. Disperse dyes are engineered colors constituting the second biggest area in the textile dyeing sector. Disperse dyes are additionally utilized for the colouration of nylon, where their principle ascribe is their capacity to cover Barré abandons, however, they're to some degree constrained speed properties to confine their utilization to shades of pale and medium profundities. Disperse dyes are additionally utilized for coloring acrylic fibres, on which they have good light speed, yet their utilization is limited to pale shades on account of their restricted develop properties. These dyes were observed to be especially helpful in the printing of cellulose mixes in connection with a regular reactive color. Disperse dye exist in the color shower as a suspension or scattering of minute particles, with just a little sum in genuine arrangement. Disperse dyes may be classified based on their rate of dyeing as low, medium and high energy dyeing [12].

5.2.4 Sulphur Dye

These dyes are water insoluble, non-ionic and sulphur linkage possessing dyes. It has been estimated that total sulphur dye constitutes about 30% of total consumption of all dyes for use with cotton. The preferred anion is delivered by diminishing and solubilising at a high temperature, and it has a proclivity for cellulose with a direct strike rate. Sodium sulphide enhances the reduction and solubilization at faster rate. It disjoins sulfur linkages and separates color particles into thiols, further to sodium salts of thiol which are dissolvable in water to sodium salts of thiol which are dissolvable in water and significance towards cellulose. Sulphur dyes can be classified according to their solubility as Conventional, Solubilised and Pre-reduced sulphur dyes. Sulphur dyes are available in powder, liquid and granular form. It is an eco-friendly dye when used with non-polluting agents [13].

5.2.5 Vat Dye

Vat dyes are water insoluble dyes differing from other synthetic dyes such as acid, basic, sulphur, reactive, etc. Vat implies 'vessels'. The vat colors are naturally acquired shading materials from the antiquated time and kept into wooden vat and make solvent in vat by the procedure of fermentation so called vat colors. Cotton (cellulosic fibres) is the most suitable material that can be dyed using this method. In contrast, wool cannot be dyed using vat dye. Vat dyes are most commonly derived from indigo dyes only. Vatting occurs at alkaline environment. This method is implemented because of its high stability, good fastness property, etc. The major limitations include its cost and tendency to cause skin diseases [14].

6 Preparation and Pre-treatment of Materials Before Dyeing

Preparation/Pre-treatment Processes used to expel pollutions from fabric strands to make it suitable for dyeing or printing. Usually the material to be dyed i.e., raw material contains unwanted impurities which may be natural impurities (wax, oil, fat), impurities from outside environment (dust, leaf parts, oils from machines) or those impurities that are additionally incorporated during the process (auxiliary agents). Regular strands and engineered filaments contain essential polluting influences that are contained normally and auxiliary contaminations that are included during turning, sewing, and weaving forms. Likewise, oils, sizes, and other remote issue are included for enhanced spinnability (in yarn make) or weavaability (in texture maker). Textile material pre-treatment is the arrangement of cleaning operations. All debasements which cause unfavourable impact during colouring and printing are expelled in pre-treatment process. Textile materials have an assortment of polluting influences in grey state or promptly during assembling. There is a chance for textile materials to be contaminated with impurities during production. Every single such polluting influence or outside issue is to be expelled from textile materials for better shading (colouring or printing) or to make them attractive in the white pattern [15]. Pre-treatment is a non-added esteem phase of the dyeing procedure furthermore, the pre-treatment phase is regularly not improved. As often as possible overabundance amounts of chemicals, agents and utilities such as water, steam, power and time are required as a part of preliminary procedures. This outcome in a high remainder of deposits like cotton polluting influences that will impact both dyeing ability and the colouring framework or will require long multi-organize strategies. The pre-treatment process must adjust the prerequisites of the colouring and completing stages and the planned end-utilization of the material. Such steps, called preliminary procedures, depend for the most part on two factors to be specific:

- The sort, nature, and area of the pollutions introduce in the fiber to be handled.
- The fiber properties, for example, salt corrosive sensitivities, protection from different chemicals, and so on.

7 Dyeing Technologies

Dyeing can be done in any phase of the preparation step. Thus, the dye can be connected to fibres, yarns, texture, or the last final stage of clothing. The real supporters of ecological maintainability in production and colouring are water and vitality, the nature and amounts of dyes and chemicals utilized, and the releases to air and stream. The quantum of each of these supporters relies upon the fibre sort, the colour, and the related compound utilization [16]. The sort and model of the hardware utilized for handling the texture additionally affect dye use proficiency and synthetic or chemical utilization.

As of late, many researches have been made to enhance different segments of dyeing, and new advancements have been created to lessen damaging fibre, diminish vitality utilization and increment profitability. New sustainable advancements that enhance the dye ability are ultrasound, ozone, plasma, ultraviolet, gamma illumination, laser, microwave, particle implantation, and supercritical carbon dioxide.

7.1 Exhaust Dyeing

Exhaust dyeing process is also termed as batch, discontinuous, direct or coordinate dyeing. Direct dyeing involves the direct application of dye to fabric without the help of any fixing agents. In this strategy, the dyestuff is either matured or synthetically decreased before being used. The coordinate dyes, which are to a great extent utilized for dyeing, are water dissolvable and can be connected directly to the fibre from a fluid environment. Most different classes of manufactured dye, other than vat and sulfur colours, are additionally connected along these lines. They create full shades on cotton and cloth without mordanting and can likewise be connected to rayon, silk, and fleece. Coordinate dyes give splendid shades, however, show poor wash quickness. Different after medications are utilized to enhance the wash quickness of direct dyes, and such colours are alluded to as after treated coordinate dyes. Direct Dyes are particles that hold fast to the texture atoms without assistance from different chemicals.

During the process of exhaust dyeing with reactive dyes, the primary or initial period of dyeing or colouring is completed under neutral pH conditions to permit colour depletion and dissemination. Chemicals like sodium chloride or sulfate are utilized to advance depletion. The temperature of the dye bath is step by step expanded to help the movement of dye into the fibre strands, and to help uniform movement. Obsession of the colour is then accomplished by adding a proper antacid to the dye bath. The response period of the colouring takes place for more than 30–60 min, with colouring temperatures in the scope of 30–90 °C, contingent upon the kind of groups involved in the reaction and their activity. The obsession procedure brings about extra dye exchange to the fibre, which is regularly alluded to a second stage or secondary exhaustion process. The auxiliary colour fatigue and dye–fibre response proceeds until the point when no further colour is taken by the fibre. Once the dye-fibre reaction stage is completed, the texture contains three structures of the dye namely (i) dye bonded covalently to the fibre (ii) dye adsorbed but not reacted (iii) dye that is hydrolyzed. The unreacted and hydrolyzed colours are not fixed. This causes poor wash-quickness and requires expulsion of the entire unfixed colour in a wash-off stride. Leftover soluble base and ingested salts should likewise be evacuated. The wash-off process requires huge amounts of water. Roughly 75% of the water utilized as a part of dyeing process is required during final washing stage. Only minimal salt concentration is required for exhaust dyeing process. The water quantity needed for this process is also low. Due to this the hydrolysis of reactive dye can be reduced resulting in high dye yield [17].

Mechanism of Direct Dyeing

The mechanism of dyeing for the utilization of direct colors to cellulose fiber includes the adsorption, dispersion (diffusion) and movement over fiber. A few components influence the coloring system; most essential is the structure of cellulose fiber, morphology and the utilization of electrolytes. At the point when the cellulosic fiber is drenched into the water the amorphous part of the fiber swells to deliver little pores at the range of 20–100 A units-the littler size color atoms diffuse into the fiber structure through these pores. The expansion of electrolytes such as Sodium chloride, sodium sulfate helps the dissemination and depletion of direct color anionic by killing the negative surface charge of cellulosic filaments. At that point, the color anions end up noticeably fastened to the cellulosic fiber through hydrogen holding and van der Waals forces.

7.2 Continuous Dyeing

Continuous dyeing processes are especially essential for cotton and recovered cellulose filaments since materials delivered from these strands are regularly colored in bigger amounts in one and a similar shade. The material substrates are fed ceaselessly into a colour range. The velocities can differ between 50 and 250 m for every moment. As per Industry gauges, Continuous coloring is a prevalent coloring technique and records for around 60% of aggregate items that are colored [18].

A Continuous coloring process normally comprises the accompanying: Color application, color obsession with heat or chemicals and lastly washing. Continuous coloring has been observed to be most reasonable for woven textures. For the most part continuous dyeing extents are intended for coloring mixes of polyester and cotton. Padding assumes a key part in the operation of continuous dyeing. Nylon floor coverings are additionally colored in continuous procedures, however the plan ranges for them is dissimilar to that for level textures. Twists are additionally colored in this process. Other types of such twist dyeing are long chain twist coloring and slasher coloring utilizing indigo. An outline of continuous dyeing process is shown in Fig. 4.

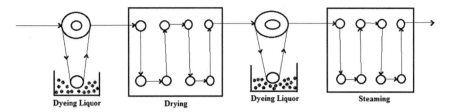

Fig. 4 Continuous dyeing process

A continuous color range has been discovered helpful and monetarily reasonable for coloring long keeps running of a given shade. One imperative factor that isolates continuous coloring from batch and other coloring is the tolerance factor for shading variety. That is more for continuous dyeing when contrasted with batch coloring [19]. This is so a result of two reasons (a) the process rate (b) nearness of an extensive number of process factors which influences color application. Few continuous dyeing processes includes pad-stream process, pad-dry process and thermosol process

(i) **Pad-Steam**

Pad Steam dyeing is a procedure of steady coloring in which the texture in open width is padded with dyestuff and is then steamed. Pad steam coloring with chosen reactive dye is the easiest and most prudent process for creating light to medium shade profundities on cotton clothing textures. The texture is cushioned with a receptive color arrangement containing inorganic electrolyte for dissemination and resulting levelness of the colors into the fiber and the fitting soluble base to start the dye–fiber response took after by 60–90 s steaming for color obsession. Steamers utilized for obsession transport the texture, as profound circles, on movable rollers. They give anaerobic condition by methods for a water leave seal. Every scope of colors requires a specific arrangement of steaming conditions.

Early steam obsession forms for the receptive coloring of cotton took after the succession: pad (with the unbiased color arrangement, salt and alkali), dry, steam, and wash. This grouping is regularly alluded to as the pad dry-compound pad steam process. This procedure arrangement resembles the steady coloring of cotton textures. The fabric and the neutral reactive dye are first padded together followed by drying using hot chamber. During the reaction, dilute sodium hydroxide solution is added for padding and finally steamed for the dye and fibre reaction to complete. In late decades, less complex and all the more naturally reasonable coloring arrangements, for example, pad dry-steam and pad steam have been presented by the real apparatus producers in a joint effort with dye providers.

(ii) **Pad-Dry**

Pad dry dyeing with reactive dyes, as its name suggests, includes padding, intermediate drying followed by color obsession by heating at high temperatures of up to 160 °C for 60–120 s. The significant factor in the pad-dry process depends on the determination of dye depending on their reactivity. The procedure is finished by washing off. The padding arrangement contains the color, a soluble base, urea, and an antimigrant. An antimigrant is utilized as a part of the padding fluid to stay away from dye movement, which causes shade changes on the two sides of the texture. The hot, dry condition gives an appropriate medium for color dissemination inside the fiber. It likewise enhances color obsession and shading yield. Urea is additionally utilized as a part of pad dry dyeing process to enhance color dissolvability. Urea is evacuated in the washing-off process, however, due to the presence of high nitrogen content, contamination and pollution problems may arise. Intermediate drying is a basic advance in any constant coloring process. Intemperate dissipation and color relocation are the key issues.

The texture is first padded in a padder with receptive color in the nearness of a soluble base. The padded texture is then gone through a pressing roller into a dryer. At the time of drying because of higher temperature obsession of color in fiber increases and in the meantime water is expelled by vanishing. The unfixed dye is washed away in the dyeing machine after drying process.

(iii) **Thermosol**

Thermosol technique is persistent strategies for coloring with scattering dye. Here coloring is performed at high temperature like 180–220 °C in a closed vessel. The time of coloring ought to be kept up deliberately to get required shade and to hold required texture quality. This coloring procedure is created by Du Pont Corporation in 1949. Here at adequate temperature the filaments relax and their inside structure is opened, polymer macromolecules vibrate enthusiastically and color particles diffuse into the fiber. It requires just a couple of seconds to 1 min and temperature around 200–230 °C. The texture is first padded with color arrangement utilizing above formula in a three bowl padding ruin. At that point, the texture is dried at 100 °C temperature in the dryer. For coloring, infrared drying strategy is a perfect technique by which water vanishes from texture in vapor frame. This dispenses with the relocation of color particles. At that point, the texture is gone through thermosol unit where thermo settling is done at around 205 °C temperatures for 60–90 s relying upon the sort of fiber, color and profundity of shade. In thermosol process around 75–90% color is settled on texture. After thermo settling the unfixed colors are washed off alongside thickener and different chemicals by warm water. At that point cleanser wash or lessening cleaning is done if required. Higher yield, dye reusability, no carrier and crease formation and requirement of low energy are few advantages of thermosol dyeing process.

7.3 Semi-continuous Dyeing

In a semi-continuous dyeing process, the texture is first impregnated with the color alcohol in, what is known as a padding machine (Fig. 5). At that point, it is subjected to cluster insightful treatment in a jigger. It could likewise be put away with a

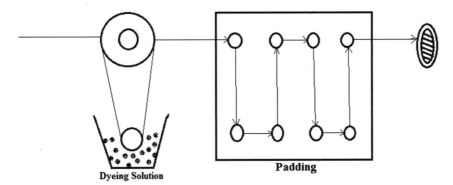

Fig. 5 Semi-continuous dyeing process

moderate pivot for a long time. In the pad batch, this treatment is done at room temperature while in pad-jig it is done at expanded temperature by utilizing a warming chamber. This results in the fixing of the colors onto the fiber. After this obsession procedure, the material in full width is completely purified and flushed in continuous clothes washers. There is just a single purpose of distinction amongst Continuous and semi-continuous coloring process is that in semi-continuous coloring, the color is connected constantly by padding [20]. The obsession and washing stays interrupted. Liquor Ratio in semi-continuous coloring is not of much significance and is not taken as a parameter.

Some of semi-continuous dyeing processes include pad-roll process, pad-jig process and pad-batch process.

(i) **Pad-batch**

Pad-Batch Dyeing is one of the generally utilized procedures for the semi-continuous coloring process. It is basically utilized as a part of the coloring of cellulosic fiber like cotton or thick (weave and woven texture) with responsive colors. Pad-batch coloring forms are the most prudent of all pad coloring forms for the responsive coloring of cotton. For little heaps of around 10,000 m, this procedure is more sparing than debilitating coloring, principally because of lower vitality necessities. This procedure additionally alluded to as cool pad-batch, includes padding the texture with a dye arrangement containing an appropriate salt framework and after that winding the texture onto a roller before grouping for 24 h. To keep away from dissipation from the uncovered edges of the roll, the texture is wrapped with material and afterward fixed with a plastic film. For color obsession at a certain temperature, the colors must have a high reactivity. After batching process, the texture is washed to expel unfixed color and leftover chemicals. This is done either on a steady washing range or on a group coloring machine. With respect to jet coloring, cool pad-batch colored textures may have an enhanced handle and cleaner surface appearance. Distinguished water and vitality reserve funds diminished utilization of colors and chemicals, and less space and work prerequisites of the pad-batch coloring process make it temperate and environmentally practical. The major advantages of Pad-batch dyeing process include adaptability, versatility, and a considerable minimising in the capital venture for hardware [21].

(ii) **Pad-Jig**

Even dyeing and good penetration can be obtained through this method. The material to be dyed is first padded with the water suspension containing vat dyes at high temperature. Dispersing agents such as sodium lignin sulfonates may be added for enhancing dyeing process. Additionally if the textile fabric does not possess good absorbing capacity, wetting agents may be added. The material after padding is arranged into rolls according to their length, thickness and fabric weight. Then the separated material is transferred into jig containing caustic soda and hydrosulfite at different pressures and high temperature depending on the type of dye used in the process.

7.4 Pigment Dyeing

Pigment dyeing is not coloring in the genuine sense as the shade sticks on the texture as a result of the binding agents. During the process of pigment dyeing, no real synthetic response happens between the dye and the texture. Rather, what happens is that the shades get situated on the texture with the assistance of binders. Pigments are not dissolvable in water and demonstrate no liking or affinity for fiber. Along these lines, regular dyestuff-based coloring conditions are not achievable for pigment dyeing. To overcome these drawbacks, another sort of colors has been detailed for use in fabric strands. These are kept up in a steady scattering in the medium of water by anionic surfactants. This sort of shade is known as pigment resin colour (PRC), essentially utilized as a part of printing.

7.5 Air-Dye Technology

Textile industry is one of the biggest utilizes of water. Ancient dyeing techniques involve the application of enormous amount of water which is then released as effluents containing hazardous chemicals thus polluting environment. In order to overcome this problem, a new technology namely Air-Dye technology has been developed involving the utilization of air instead of water for dyeing process. This technology is also called Water-free Dye Technology. Flow of air is the main idea of this technology, air acting as medium of transportation of dye to fabrics. Replacement of water with air is the promising idea in this technology with no treatment or finishing process is required. Dyeing can be done at the two sides of the fabric material. Dyeing is done inside the fabric material thus avoiding the problem of leaching or fading of dyes. It is an economically feasible sustainable technology with more benefits.

This process can be performed in a single dyeing machine without the involvement of other processes such as streaming, washing, etc. The major advantages of air-dye technology are

- Very low water consumption/Release of wastewater
- Lesser process time
- High energy savings
- Highly flexible and maximum colour durability is obtained

Mechanism

This method does not require water for dyeing instead this employs air to enter into fibres. In this method, the fabric is first heated and then the dye is injected directly into the fibres in the form of gas. The outcome of this technology is more beneficial than any other conventional dyeing methods such as vat dyeing, cationic dyeing, etc. The colour after dyeing process results in rich look and lasts for a longer period of time [22].

7.6   Supercritical Fluid Dyeing

Supercritical fluids are generally gases possessing the properties of both liquid and gas acting as both solute and solvent. Supercritical fluids are substances above their critical temperature and pressure. At this point the substances does not evaporate or condensate to form a gas or liquid. They possess high solute diffusivity and are low in viscous. But the transportation process in supercritical fluids is high. Carbon dioxide is the most commonly used supercritical fluid because the critical pressure and temperature can be easily obtained compared to other substances. The critical point of carbon dioxide takes place at temperature of 31.1 °C and pressure of 73.8 bar. Supercritical CO_2 is abundantly available, inert, non-flammable, non-explosive, non-toxic, and inexpensive and can be recycled [23].

Mechanism

The dye and fibre are added to the reaction vessel. The components present in a CO_2 dyeing system are CO_2 gas cylinder, pressure pump, temperature controller, vessel, heating and cooling system. The whole system is pressurized with CO_2 up to 800 Psi. Continuous stirring is done with agitation speed of 1000 rpm. Temperature of about 180 °C is maintained. Then pressure is raised to 3500 Psi and the system is maintained at these conditions for 2 h. Finally the pressure is released and dyed fibre is removed.

Advantages

The following are few advantages of using supercritical fluids for dyeing than any other conventional methods:

- No discharge of wastewater/contaminated water into environment.

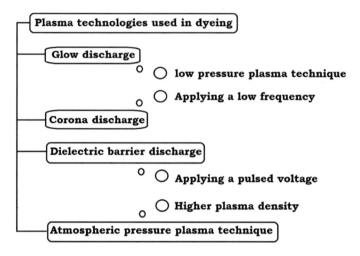

Fig. 6 Plasma technologies used in dyeing process

- CO_2 causes swelling of fibre thus enabling dyes to diffuse at faster rate.
- Energy required for dyeing process is low compared to other conventional methods.
- Drying process is not required after dyeing.
- Supercritical fluids cause no pollution, non-flammable and are nontoxic.
- Diffusion rate is comparatively higher.

7.7 Plasma Technology

Plasma is defined as an ionized gas with equal ratio of positive and negative charges under extreme condition of pressure and temperature. Plasma are formed when a substance at gaseous state maintained at high energy results in release of outer electron from an atom and this released electron becomes free negative charge and atom becomes free positive charge. The ionization of plasma gas can be induced for dyeing process using various methodologies such as Dielectric-barrier discharge, Atmospheric pressure plasma technique, Corona discharge, Glow discharge, etc. (Fig. 6).

Principle of Plasma Technology

Plasma is composed of radicals, ions, electrons, ultraviolet radiations and other molecules that react with the substrate. Plasma technology is mainly used for inducing surface modifications and also for enhancing the property of textile materials for increasing dyeing rates, for colour improvement, diffusion and adhesion of coated dyes. The textile material to bicoloured is placed inside the chamber and plasma is incited. The particles gets generated and then interacts with the surface of the textile material. A thin nanometre sized film is formed on the surface of the material and the surface is structured with functional groups [24].

Advantages of Plasma Technology

- Chemical and Water discharge is less.
- Colour obtained is bright and durable.
- This method alters the surface of fibre than modifying inside the material.
- Effect on environment is very less.

7.8 Foam Dyeing

Foam is defined as the dispersion of gas in liquid. Gas used is mostly an inert gas and liquid used is water. Foam is the key factor in foam dyeing process. Foam are formed using foaming agents and usually foam is mainly obtained from aqueous solution which is then spread on the textile material. These agents must produce foam instantly, should not get affected by temperature, quick wetting process and ability to stabilize itself. Foam may be of dispersion foam or condensation foam. Dispersion foam is mixing of gas with the liquid while condensation foam is producing gas within the liquid physically or chemically.

Mechanism

The materials required for foam dyeing are synthrapol for removing dirt and other contaminants from textile material, foam cream, soda ash and dye to be fixed. The fabric is first washed with synthrapol and soaked in a solution containing soda ash. Then the foam cream is mixed well to form a creamy texture and poured into the chamber along with the dye to be coated. The cleaned fabric is placed on the foam and soaked for a certain time interval at elevated temperature. During this incubation, the dye gets attached to the fabric surface with higher stability.

Advantages

- Fixation of dye into fibre can be improved.
- Diffusion of dye into fibre can be enhanced.
- Stability of the fibre dyed obtained is high.
- Outcome is more in short time duration.
- Waste generation is less and energy saving process.

7.9 Microwave Technology

Thermal and dielectric properties play a major role in microwave dyeing. This method is used for dyeing small quantities of textile materials in microwave with the help of Procion dyes. Dielectric property indicates intrinsic electrical properties affecting dyeing process through bipolar rotation of the dye and has impact on the microwave field in the dipoles. The aqueous dye solution has polar components in the high frequency microwave field in the range of 2450 MHz. It affects the vibrational energy of the dye molecules as well as water molecules. Resistance heating is provided through ionic conduction. Heat can be transferred through conduction, convection or radiation. Due to ionic acceleration in the dye solution, collision takes place between the dye molecules and the fibre molecules. Dye penetration takes place with the help of mordant. Mordant also helps in deeper penetration of dyes into the fabrics [25].

Mechanism

The fabric material is washed prior to dyeing. Hot water is added to the microwave container containing fabric material and dye powder is added to it. The container is closed and covered properly. Then the container is placed inside the microwave and treated at high temperature for few minutes. After that dye solution is added again and the process is repeated. Then the container is removed and cooled. The dye gets absorbed to the fibre leaving the cloudy water. Then water is filtered and the fabric is dried in shade. The main drawback in microwave dyeing technology is that uniform dyeing cannot be obtained and the depth of dyeing is also not even. Colouring of fabric material occurs but most of the dye stays in water only and is washed out during rinsing process.

7.10   Ultrasonic Wave Dyeing Technology

Ultrasound is a vibrating sound pressure wave having a frequency greater than the limit of adults hearing range. Normally the frequency of ultrasound ranges between 20 kHz and 500 MHz. The main equipment used for ultrasonic are Generator, Converter and Transducer. Generator is used for converting alternating current to high frequency electrical energy. The converted electrical energy is then fed to the transducer and converted into mechanical vibration. The main function of transducer is to vibrate longitudinally and transmit the waves into the liquid medium. Propagation of these waves results in cavitation. The factors affecting cavitation and bubble breakage are gas and solvent properties, pressure, temperature and frequency of sound waves [26].

Mechanism

Cavitation occurs when ultrasonic waves are absorbed into the liquid system. This results in release of entrapped gases in the liquid medium such as the textile material or dye solution. The effect of ultrasound technology on dyeing process can be explained in three methods:

- Dispersion: Breaking of micelles and high molecular weight compounds to form uniform dispersion in the dye solution.
- Degassing: Release of entrapped gases from the fibre capillaries.
- Diffusion: Penetration of dye into the fibre material. Interaction occurs between the dye and fibre resulting in bond formation.

Advantages

- Energy saving process and temperature required is also low.
- Operating time and chemical usage is also less.
- Product quality can be improved.
- This method is suitable for water insoluble to hydrophobic dyes.
- It requires less processing cost.

Disadvantage

The main drawback of using ultrasonic wave technology in dyeing process is difficulty in producing uniform ultrasound waves and high intensity in a large vessel.

7.11 Ozone Technology for Dyeing

Ozone is a naturally occurring gas that has both beneficial and hazardous effect on the environment. It is mostly present in the stratosphere and protects the earth from harmful ultraviolet radiation entering it. It is a pungent smelling gas. Ozone gas can also be produced artificially by various methods such as Electrolysis, Corona discharge and UV radiation. Ozone is a strong oxidising pungent smelling gas. Ozone gas is helpful in surface modification and improving fibre durability through a process termed ozonation. The dyeing ability through ozonation process depends on factors such as pH, temperature, water level and ozone dosage level [27].

7.12 Bio-based Dyeing Technology

Conventional dyeing techniques have negative impact on environment though they result in rich colourful products. The presence of toxic chemicals, heavy metals and other hazardous substances affect humans who wear it. To overcome these issues, new technique namely "bio-based dyeing" has been developed with more benefits such as safe, eco-friendly, durable and also cost-effective. These dyes are also known as natural dyes. Plants, animals and microbes are used for this type of dyeing process. Compared to plants and animals, microbe-based dyeing is more effective with high efficiency. Downstream processing can be eliminated using bio-based dyeing technique. The dye is in liquid state and dyeing can be done in batch or continuous mode. The dyeing process depends on several parameters such as type of textile material, production conditions, requirement of product quality, etc.

7.13   Electrochemical Dyeing

The synthetic dyes such as vat and sulfur are insoluble in water; for their application, it is important to change them into the water-solvent nature utilizing appropriate reducing agents and soluble base. The procedure utilizes an electric current rather than chemical substance reducing operators, giving it variously specialized, economic and biological advantages. The utilized color solution can't be reused due to the fact that the reducing energy of these chemicals can't be recovered. The transfer of the dye solution and the effluent water cause different issues due to the non-ecofriendly nature of the deteriorated compounds. Electrochemical dyeing is still in the research center stage yet could turn into the coloring procedure without bounds of the vat, indigo and sulfur colors. Electrochemical dyeing in which synthetic reducing agents and effluent polluting substances are distributed from inside and out. There are two techniques by which electrochemical dyeing can be done: direct or coordinate electrochemical coloring and indirect electrochemical coloring.

(i) **Coordinate electrochemical colouring**

In occurrence of direct electrochemical dyeing procedure, natural dyestuff has been specifically lessened by dye and cathode contact. The dye solution is reduced by utilizing the regular reducing agents and afterwards entire color reduction is accomplished by an electrochemical procedure for completing reduction process which encourages the enhanced steadiness of the reduced color.

(ii) **Indirect electrochemical colouring**

Here, the dye isn't decreased at the terminal. A reducing agent is included that decreases the color in a regular way thus gets oxidized after color reduction. The oxidized reducing operator is therefore decreased at the cathode surface, which is then further accessible for color reduction. This cycle is consistently done at the time of coloring operation. In electrochemistry, the chemical agent undergoing both reduction and oxidation process is known as reversible redox framework and is known as an arbiter. Consequently, the color reduction does not happen because of direct contact of dyestuff with the cathode, as indirect electrochemical reduction; however, it happens through the arbiter which gets more decreased because of the contact with the cathode.

8   Dyeing Machines

Dyeing process is finished by various sorts of coloring machine. The machine which is accustomed to coloring or shading of materials like yarn, texture, pieces of clothing or some other materials is called dyeing machine. Dyeing machines come in all shapes and sizes to suit the different structures and amounts of textile materials. In reality, the gadget is utilized by various sectors for dyeing process. Different sorts of coloring apparatus are utilized to color the material materials. The dyeing machine producing sector must pose necessary actions for the development and generation of modern dyeing machinery through various research and development (R&D) in order to overcome the market needs. The main goal of developing machines must meet the following criteria

- Labour requirement must be reduced.
- Increasing the yield and productivity.
- Easy operating conditions.
- Minimising the cost and effluents thus controlling pollution.
- Energy and water saving system.

Presently, desired quality, efficiency and productivity can be achieved only through electronically controlled process conditions. Nowadays new technologies have been introduced during the development of dyeing machines to overcome the needs in the market. Basically dyeing machines can be classified based on their operation as open dyeing machine and closed dyeing machine. The following are few dyeing machines commercially used for dyeing process [28].

8.1 Jigger Dyeing Machine

A jigger is a dyeing machine in which texture in open width is exchanged over and over starting with one roller then onto the next and sits back through a dye liqour of moderately little volume. Jigg or jigger coloring machine is one of the most established coloring machines utilized for fabric dyeing operations. Jigger machine is appropriate for coloring of woven textures, up to boiling temperature with no wrinkling. Jigg applies extensive the long way strain on the texture and is more appropriate for the coloring of woven than weaved textures. The fabric is passed through the roller through the dye liquor, and for another roller is given an intensive treatment in the bath. The treatment in this open vat can be done again and again until the point when the expected shade appears on the fabric. The coloring procedure on jigger is viewed as a progression of discontinuous padding operation took after by abide periods on the fundamental roller, amid which the coloring move and dissemination make put. The significant disadvantage in jigger coloring machine incorporates because of a low solution proportion washing-off procedure is very troublesome, minimal mechanical activity in a jig machine and it is less appropriate where overwhelming scouring is required before dyeing.

Principle

The dyeing procedure on jigger is viewed as a progression of irregular padding operation took after by harping time on the fundamental roller, amid which the coloring move and dissemination occur. The variables controlling the rate of color absorption are:

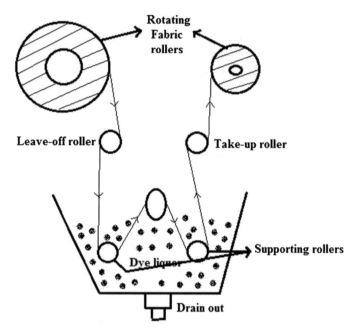

Fig. 7 Working of jigger dyeing machine

- The measure of interstitial color solution held in the interval of the textured weave.
- The fatigue of the interstitial alcohol in the stay time frame between progressive submersions.
- The level of exchange of solution at the time of one submersion.

In dyeing on jigger machines the material spins on two fundamental rollers, the open-width texture goes from one roller through the dye solution at the base of the machine and after that onto a determined take-up roller on the opposite side. At the point when all the texture has gone through the solution, the pathway is switched. Every section is called an end. Coloring dependably includes a considerable number of finishes. The dye solution has at least one guide rollers, around which the fabric ventures, and amid this submersion accomplishes the necessary contact with the dye solution. During this entry, the texture grabs satisfactory amount of dye, the abundance of which is depleted out yet a decent amount is held in the texture. During revolution of rollers, this adsorbed color infiltrates and diffuse into the texture. The genuine coloring happens, not in the dye solution but rather when the material is on the rollers since just a little length of texture is in the dye liquor and a significant part is on the rollers (Fig. 7). The rate of fabric at the time of immersion in dye solution has an almost no impact on the level of shade delivered.

Advantages

- The fabric material can be dyes in open full width forms.

- Loss of steam and heat energy is low compared to other dyeing machines.
- Chemical loss is also less since the ratio of textile material and dye liquor ratio is 1:3.

Limitations

- It applies a great deal of strain in the twist direction and as a result of this ordinarily woolen, weaved textures; silk, etc. are not colored in jigger dyeing machine.

8.2 Winch Dyeing Machine

The winch is the most established known simple and minimal effort rope coloring machine with static dyeing liquor and material moved by means of a winch reel or roller. This machine applies just a low pressure on the texture, lower than that applied by a jigger, and subsequently was viewed as a perfect coloring machine for fragile and strain touchy textures, for example, thick rayon and the weaved materials. The machine works at a greatest temperature of 95–98 °C. The dye liquor solution is applied to the fabric materials in a high ratio of about 1:20. Winch coloring machines are low-cost operators that are easy to work and continue, yet flexible in application demonstrating significant for preparing, washing and additionally the coloring phase itself. In all winch coloring machines, a progression of texture ropes of equivalent length is submerged in the color shower yet part of each rope is assumed to control two reels or the winch itself. The rope of texture is flowed through the dye liquor being pulled up and over the winch over the span of the colouring operation. Dye solution and helping agents might be dosed physically or consequently as per the formula technique.

Principle

The fundamental rule of all winch colouring machines is to have various circles or ropes of the texture in the dye solution, these ropes are of equivalent length, which is generally submerged in the solution. The upper piece of each rope keeps running more than two reels which are scaled over dye solution. At the front of the machine, over the highest point of the dye solution is a little reel, which is called manoeuvre or fly roller. At the back of winch, a tank called the winch wheel, which pulls the textured rope from the dye solution over the manoeuvre reel for dropping in the dye liquor for submersion. From the dropped area, the textured rope goes back to be lifted and encouraged to winch wheel (Fig. 8). The colouring process on winch dyeing machines depends on higher textile material to dye liquor solution ratio (1:20) as contrasted and other colouring machines. The procedure is led with low to no strain. The aggregate dyeing time is long-lasting when contrasted with different dyeing machines.

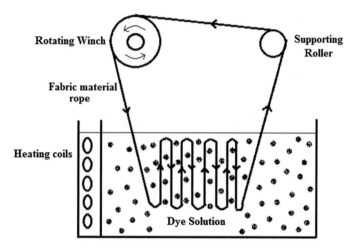

Fig. 8 Working of winch dyeing machine

Advantages

- Simple in operation and construction, clean and smooth textured fabric can be obtained.
- Winch dyeing machines are more suitable for wet processing operations from pre-treating processes like de-sizing till whitening.
- Strain applied in winch is less compared to jigger dyeing machine.

Limitations

- Dyeing of continuous full length fabrics is not possible in a single batch process.
- The texture is handled in rope frame which may prompt wrinkle marks, especially in overwhelming, woven, etc.

9 Future Trends in Dyeing

There is little uncertainty that the refinement of colors, process ways, machine innovation reusing and effluent treatment that has brought about noteworthy enhancements to maintainability over past decades will proceed. Some of these advancements will be driven by the consistent interest for cost diminishments. For decades, water has been the most vital component in textile dyeing. Today, it constitutes an undeniably costly medium. Effluent discharge is the major problem in all dyeing processes. The substitution of dye solution with air serves as a medium of transport in jet dyeing machines resulting in a huge reduction of chemical and water usage. The dampness immersed air flow guarantees the uniform conveyance of temperature on the texture and in the machine, constituting an essential for even and reproducible coloring.

Accordingly, the danger of draft or strain is insignificant, which is especially beneficial with respect to the completing process. The Airflow innovative technology speaks to the consolidated outcome of the whole scope of potential outcomes.

The stability and efficiency of dyeing process can be improved by the following steps:

1. Advancement of engineered colors with enhanced sciences.
2. The utilization of ecologically more secure chemicals in dye solutions.
3. Fibres can be modified chemically before dyeing process.

R&D has been focused on improving coloring execution through changes to the nature and number of the groups on the colour molecule. The degree of dye–fibre response and a definitive release of unfixed colour changes generally with the kind of receptive group and the coloring innovation utilized. Enhanced fixation levels and consequently bring down color expenses and release amounts have been accomplished through utilizing two distinctive useful gatherings on the color molecule. The utilization of financially accessible bi-or polyfunctional receptive colors has been suggested as the best accessible system for expanding the color fixation proficiency. Biodegradable dye solution chemicals offer a fascinating other option to inorganic salts. The significance is in reducing effluent levels. Natural surfactants help in reducing these effluent levels.

The oxidation and reduction forms in vat and sulfur dyeing create effluent contamination. Sodium dithionite, as the overwhelming decreasing agent, creates a lot of sodium sulfate (TDS) and ecologically unwanted sulfite and thiosulfate as an outcome. Biotechnological processes utilizing catalysts for reduction and oxidation reaction do offer some potential for its application in the textile industry.

Synthetic or chemical alteration of cotton to enhance coloring with immediate, receptive, sulfur, or vat colors is a rising technology. Research has concentrated on the acquaintance of cationic functional groups with fiber. The modifications are enhanced by processing cotton with low atomic weight cationic chemicals. Thus, higher fixation productivity and decreased utilization of salt can be accomplished [29].

10   Conclusion

The textile industry is a chemical oriented sector. Factors, for example, scale development, expanded time to advertise, the division of work, and utilization designs have profoundly impacted how a textile product is created. Taking into account both the quantity and the structure of chemicals in textile effluents, for example, the availability of colours, salts, added substances, cleansers, and surfactants, the textile sector is evaluated as the most contaminating and polluting sector among other industries. Research is going on to overcome the issue of environmental pollution and contamination. Plasma and supercritical carbon dioxide have been investigated as innovative approaches to wipe out the utilization of water in dyeing. Every administration is in charge of thinking about the pollutant-free methodologies beginning from the materials chosen to supply of the finished goods for trade. To reduce the release of toxic contaminants from the textile industry, it is important to minimize the utilization of colors, supporting chemicals and water. The subject and the issues included are quite differed and complex. Dyeing cotton clothing textures with responsive colors speak to the biggest volume in all clothing coloring. It is additionally the biggest supporter of natural contamination.

References

1. McCarthy, B. J. (2016). 1—An overview of technical textiles sector. In *Handbook of technical textiles* (2nd ed., pp. 1–20).
2. Babu, B. R., Parande, A. K., Raghu, S., & Kumar, T. P. (2007). Cotton textile processing: Waste generation and effluent treatment. *Journal of Cotton Science, 11,* 141–153.
3. Smith, B. (2003). Wastes from textile processing. In A. L. Andrady (Ed.), *Plastics and the environment* (1st ed.). Hoboken, USA: Wiley.
4. Tandon, N., & Reddy, E. E. (2013). A study on emerging trends in textile industry in India. *International Journal of Advancements in Research & Technology, 2*(7).
5. Kumar, P. S., & Suganya, S. (2017) 1—Introduction to sustainable fibres and textiles. In *Sustainable fibres and textiles* (pp. 1–18).
6. Chakraborty, J. N. (2010). 1—Introduction to dyeing of textiles. In *Fundamentals and practises in coloration of textiles* (pp. 1–10).
7. Holme, I. (2016). 9—Coloration of technical textiles. In *Handbook of technical textiles* (6th ed., pp. 231–284).
8. Shahid, M., Islam, S., & Mohammad, F. (2013). Recent advancements in natural dye applications: a review. *Journal of Cleaner Production, 53,* 310–331.
9. Forgacs, E., Cserhati, T., & Oros, G. (2004). Removal of synthetic dyes from wastewaters: a review. *Environment International, 30*(7), 953–971.
10. Enaud, E., Trovaslet, M., Bruyneel, F., et al. (2010). A novel azoanthraquinone dye made through innovative enzymatic process. *Dyes and Pigments, 85,* 99–108.
11. Ingamells, W. (1993). *Colour for textiles: A user's handbook*. England: Society of Dyers and Colourists.
12. Fryberg, M. (2005). Dyes for ink-jet printing. *Review of Progress in Coloration, 35,* 1–30.
13. Jimenez, M., & Estape, N. (2003). New dyeing process with sulphur dyes. *Melliand Textilberichte, 84*(9), 753–755.
14. Thetford, D., & Chorlton, A. P. (2004). Investigation of vat dyes as potential high performance pigments. *Dyes and Pigments, 61*(1), 49–62.
15. Robinson, T., McMullan, G., Marchant, R., & Nigam, P. (1997). Remediation of dyes in textile effluent: A critical review on current treatment technologies with a proposed alternative. *Colorage, 44,* 247–255.
16. Varadarajan, G., & Venkatachalam, P. (2015). Sustainable textile dyeing processes. *Environmental Chemistry Letters*. https://doi.org/10.1007/s10311-015-0533-3.
17. Shukla, S. R. (2007). Pollution abatement and waste minimisation in textile dyeing. In R. M. Christie (Ed.), *Environmental aspects of textile dyeing* (1st ed.). Manchester, England: Woodhead Publishing Ltd.
18. Shore, J. (1979). Continuous dyeing. *Review of Progress in Coloration, 10,* 33–49.
19. Huanga, T., Cui, H., Yang, D., et al. (2017). Continuous dyeing processes for zipper tape in supercritical carbon dioxide. *Journal of Cleaner Production, 158,* 95–100.
20. Nair, G. P., & Pandian, S. P. (2008). Spotlight on textile machinery: A buyer's guide to textile processing machinery: Semi-continuous and continuous open-width fabric dyeing machines. *Colourage Supplement: Machinery and Process Focus, 55*(11), 12–24.

21. Teli, M. D. (1997). New developments in dyeing control process. *Indian Journal of Fibre & Textile Research, 21,* 41–49.
22. http://www.textileworld.com/Issues/2010/MarchApril/Features/Recent_Developments_In_ Dyeing.
23. Chankraborty, J. N. (2010). 27—Dyeing in super-critical carbon dioxide. In *Fundamentals and practices in coloration of textiles* (pp. 299–306).
24. Lakshmanan, S. O., & Raghavendran, G. (2017). 9—Low water-consumption technologies for textile production. In *Sustainable fibres and textiles* (pp. 243–265).
25. Buyukakinci, B. Y. (2012). Usage of microwave energy in Turkish textile production sector. *Energy Procedia, 14,* 424–431.
26. Kamel, M. M., El-Shishtawy, R. M., Tussef, B. M., & Mashaly, H. (2005). Ultrasonic assisted dyeing III. Dyeing of wool with lac as a natural dye. *Dyes and Pigments, 65*(2), 103–110.
27. Atav, R., & Yurdakul, A. (2011). Effect of ozonation process on the dyeability of mohair fibres. *Coloration Technology, 127*(3), 159–166.
28. Nair, G. P. (2011). 8—Methods and machinery for the dyeing process. In *Handbook of textile and industrial dyeing. Principles, processes and types of dyes* (pp. 245–300). England: Woodhead Publishing series in textiles.
29. Khatri, A., & White, M. (2015). 5—Sustainable dyeing technologies. In Sustainable apparel production, processing and recycling (pp. 135–160). England: Woodhead Publications, Elsevier.

Chapter 2
Eco-friendly Production Methods in Textile Wet Processes

Seyda Eyupoglu and Nigar Merdan

Abstract Sustainable', 'economical', and 'eco-friendly' production has recently become important issues in textile manufacturing processes. In the world, textile conventional production industry is one of the major industries which cause environmental pollutions. During textile wet process, great deals of wastes are leaved in air, soil and, especially water. Due to these wastes, all species in the ecosystem are negatively affected. In order to manufacture a ton of textile, approximately 230–270 tons water is used. After the textile production, the water is undertaken with heavy chemicals and this waste water is leaved in environment. In textile production industry, there are two efficient methods to decrease the environmental pollution. Constructed of large and highly effective effluent treatment plants is a method to reduce the amount of wastes. The other method is the use of natural raw materials and ecological production methods. Recently, researchers have been seeking for ecological, sustainable, and biodegradable natural raw materials alternatively synthetic raw materials. Especially, natural textile raw materials have been begun to use acceleratingly in compration with synthetic raw materials in textile industry. Furthermore, new natural fibers have been obtained from different source and the use of these fibers has searched in textile industry. In textile wet process, especially, waste water with heavy chemicals load is a major problem. In order to eliminate the negative effect of waste water, researchers have been searching for solutions. In literature, coating, microencapsulation, plasma applications, using of ultrasonic and microwave energy, using of supercritical carbon dioxide and ozone treatment are described as some of the eco-friendly process in textile wet industry. In this study, some eco-friendly production methods in textile wet industry were investigated, separately. Furthermore, the advantages of new production methods were searched.

Keywords Textile wet process · Eco-friendly production · Waste water

S. Eyupoglu (✉) · N. Merdan
Department of Fashion and Textile Design, Architecture and Design Faculty, Istanbul Commerce University, 34840 Kucukyali, Istanbul, Turkey
e-mail: scanbolat@ticaret.edu.tr

© Springer Nature Singapore Pte Ltd. 2018   31
S. S. Muthu (ed.), *Sustainable Innovations in Textile Chemical Processes*,
Textile Science and Clothing Technology, https://doi.org/10.1007/978-981-10-8491-1_2

1 Introduction

In human history, the most important equipment has been cloth after the nutritional requirement. For this reason, the history of textile fibers dates to B.C. 9000. Throughout these years, flax harvested in Mesopotamia, cotton cultivated in the region of Indus River, silk obtained from domesticated silk worm in North China and wool originated from West Asia have been used. The natural based textile fibers played a significant role due to their widespread use in nature and technical appropriateness for many centuries. In recent years, textile fibers have been used commonly in the areas of technical textiles because of low price, lightness, high strength properties, low thermal conductivity, accessibility, etc.

Starch derivatives can be used as alternatives to starch for desizing. These are carboxymethyl cellulose, heteropolysaccharides, polyvynilalcohol and polyacrilate. Chemical oxygen demand (COD) in starch is substantially lower than that of the other desizing materials however starch is the desizing material polluting waste waters the most due to utilization of excessive amount of starch desizing during desizing process. In case excessive loading of waste water by starch desizing material is desired, the simplest precaution would be the collection of desizing removal flottelers with starch waste in a special sedimentation pool by a separate canalization. Since the majority of these starch wastes will sediment, COD and biological oxygen demand (BOD) values of water overflowing at the outlet will drop substantially.

Due to increasing industrialization, environmental pollution has arisen as a global issue. In literature, it is registered that more than 700,000 tons of dyes are manufactured. However, 280,000 tons of dyes are wasted due to adverse environmental impacts. Thus, textile industry is one of the industries that damages the environment with toxic wastewater. According to the World Bank, 17–20% of industrial water pollution is caused by industrial water pollution [1–3].

Clean production approach considers all possibilities that will alleviate the current pollution problem in textile wet treatment and will save water or energy. The things to be done are grouped under 4 main headings:

1. Process optimization: Less water consumption in every possible area, less chemical use, working in lower temperatures and less time loss.
2. All chemicals are environmentally friendly (use of environmentally friendly chemicals).
3. Reuse of water: By purification.
4. New technologies (transfer printing, enzymatic processes, plasma and ozone technologies, dyeing in CO_2-containing environment).

There are many conventional methods for the purity of textile wastewater such as chemical coagulation (using ferrous, and polyelectrolytes), biological treatment followed by activated carbon adsorption. However, conventional methods have a number of disadvantages, including the generation of a huge volume of sludge. In order to meet the demands of environmental standards, researchers have investigated new methods for the complete and successful disposal of textile wastewater [4].

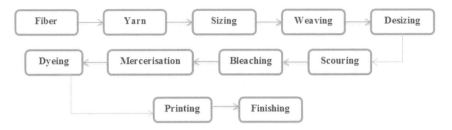

Fig. 1 Textile operations

In this study, eco-friendly production methods in textile wet industry such as ozonation, plasma applications and enzymatic treatments were investigated, separately. Furthermore, membrane filtration which is used to clean textile waste water was investigated. The advantages of new production methods were searched.

2   Textile Operations

Textile raw materials, called as fibers, are transformed to yarns, which are then turned to fabrics. These fabrics are exposed to several textile wet processes. Textile operations are shown in Fig. 1 along with some summaries of textile processes [5].

2.1   *Sizing and Desizing*

In textile industry, sizing process is known as a weaving preparatory process which is applied to warp yarns. The essential of the sizing is identified as reduction of yarn breakage and disposing the weaving machine stops. The most used sizing agents are starch, polyvinyl alcohol (PVA) and carboxymethyl cellulose (CMC). In textile process, desizing process has been used to remove sizing agents that have been applied to warp yarns during a weaving process.

Enzymatic or oxidation process transforms starch to simple water-soluble products. Desizing waste has more biological oxygen demand in the range of 300–450 ppm and pH 4–5. In degradation of starch, hydrogen peroxide can be used to convert starch into CO_2 and H_2O. Furthermore, enzymatic process can be eased by converting starch into ethanol. After the distillation of ethanol, it can be used as a fuel [5].

2.2 Bleaching

Natural color substance causes the fabric to seem like cream. In order to obtain white fabric, the natural color matter should be removed from the fabric with bleaching. In the past, hypochlorite was the most used agent but recently H_2O_2 has taken its place [5].

2.3 Mercerization

Mercerization is a chemical treatment carried out on cotton fibers to gain shine and improve dye uptake. Furthermore, mercerization increases tensile strength of cotton fibers. Basically, cotton fibers are treated with a high concentration (almost 18–24% by weight) of sodium hydroxide for usually less than 4 min. Cotton fibers are then treated with water or acid for 1–3 min under stress to neutralize the sodium hydroxide. The material later gains easy dye uptake and its absorbency increases. In order to recycle sodium hydroxide in the wash water, membrane techniques or multiple effect evaporators can be used [5].

2.4 Dyeing and Printing

Dyeing is the treatment to color textile fibers, yarns and fabrics. Being responsible to gain color, these groups can be listed as azo (–N=N–), carbonyl (–C=O), nitro (–N=O), quinoid groups and auxochrome groups like amine, carboxyl, sulphonate and hydroxyl. Among these groups, azo and anthraquinone groups are the most significant ones. Furthermore, these groups are responsible from coloring of textile waste water and contamination. Figure 2 shows the type of dyes used for different types of fibers [5].

Printing is a regional color process of textile materials. In dyeing, dye is treated in a solution form whereas dye is applied in a thick paste form in printing. Wastes of dyeing and printing have quite similar compounds.

2.5 Finishing Process

Finishing processes are called as the whole set of operations performed to improve the handle, usage properties and appearance. At the present time, finishing processes are classified as chemical finishing process and mechanical finishing process. Finishing processes are listed in Figs. 3 and 4.

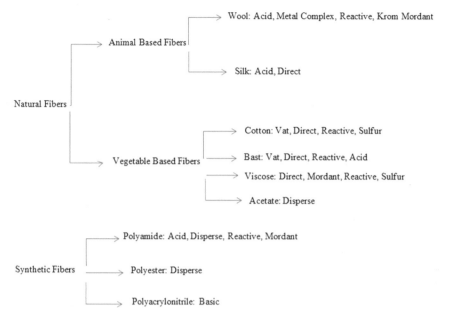

Fig. 2 The type of dyes used for different types of fibers

After the textile operations, textile wastewater is loaded with harmful chemical. In desizing process, textile wastewater is embarked with sizes, enzymes, starch and waxes. After the scouring process, scouring wastewater contains NaOH, surfactants, soaps, fats, pectin, oils, sizes and waxes. Considering the bleaching process, the wastewater includes H_2O_2, sodium silicate, organic stabilizer and alkalies. Dyeing and printing wastewater comprises dyes, color pigments, metals, salts, surfactants, alkalies, acids, ureas, formaldehyde and solvents. After the finishing process, softeners, solvents, resins and waxes appear in finishing process wastewater [5].

3 Ozonation Technology

3.1 Properties of Ozone Gas

Ozone gas was first discovered by German chemist Christian Friedrich Schönbein based upon its distinctive smell after thunderbolt. In 1839, he named this smell "ozone" which stands for the verb "smell" in Greek. In 1856, Thomas Andrews stated that ozone comprised of oxygen and then Soret discovered the relation between oxygen and ozone as shown in Formula (1) [6].

$$3O_2 \leftrightarrow 2O_3 \; \Delta H_f^0 \, at \; 1 \, atm = +248.5 \, kJ \, mol^{-1} \tag{1}$$

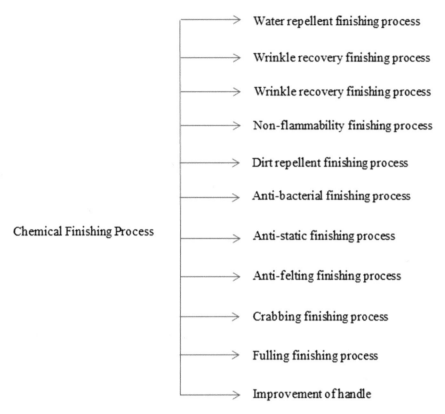

Fig. 3 Chemical finishing processes

Ozone molecule consists of three oxygen atoms which occurred at the end of the oxidation process of oxygen molecule. Ozone has high activation energy due to having a free bond. Ozone is colorless in gas form and blue in liquid form. The physical and chemical properties of ozone were given in Table 1.

Ozone is an organic molecule including three different oxygen atomic rings two of which are breathed. Oxygen is transformed to ozone because of exposure to ultraviolet rays in upper-layer of atmosphere. Due to the high weight of ozone, it comes down to the earth.

Ozone is a gas which is non-heat resistant, corrosive and is transformed to oxygen. Because of these properties, ozone is not stored or transferred and it should be produced in the environment in which it is used.

Ozone gas is used in a wide variety of industries, including iron-steel and metal, textile, chemistry, food, automotive, medical, agriculture and stockbreeding, odor and color removal in mine industry, storage, heating and cooling systems, purification of water and air, sterilization and protection of food materials.

Ozone is a powerful oxidizing agent used in some applications in swimming pools, industrial waters consisting of phenols and drinking water [7].

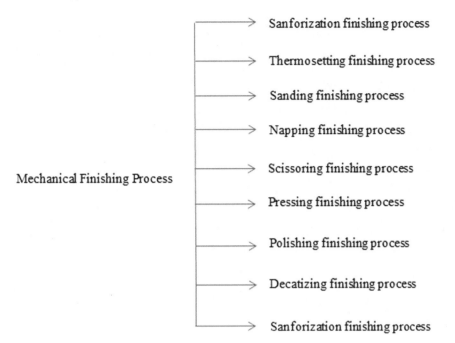

Fig. 4 Mechanical finishing processes

Fig. 5 Corona discharge ozone generator [9]

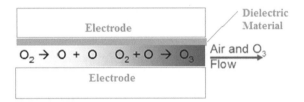

3.2 Production of Ozone Gas

Ozone gas can be produced in two different industrial methods. The first method is the use of UV at 185 nm and the second method is the use of Corona discharge. Corona discharge can be defined as an electrical discharge which is generated by the ionization of a gas surrounding a conductor that is electrically charged [8]. Production of ozone gas with Corona discharge is given in Fig. 5.

In this method, ozone is produced by supplying air or oxygen gas into the generator. In the ozone generator, oxygen or air is converted into ozone by the discharge of electric. First, primary components in air are separated into reactive atoms or radicals with the intense electric field. Then, these reactive atoms can react among themselves.

Table 1 The physical and chemical properties of ozone

Properties	Value
Moleculer formula	O_3
Moleculer weight	48.0 g/mol
Boiling temperature	$-111.9\,°C$
Melting point	$-192.7\,°C$
Critical temperature	$-12.1\,°C$
Critical pressure	5.53 MPa
Density in gas form	2144 kg m^{-3}
Redox potential	2.07 V
Density in liquid form ($-112\,°C$)	1358 kg m^{-3}
Viscosity of liquid ($-183\,°C$)	1.57×10^{-3} Pa s
Heat capacity of liquid form ($-183\,°C$ to $-145\,°C$)	1884
Heat capacity of gas form	818 J kg^{-1} K^{-1}
Heat of evaporation	15.2 kJ mol^{-1}

$$O_2 \rightarrow 2O \tag{2}$$

$$O + O_2 \rightarrow O_3 \tag{3}$$

3.3 The Usage of Ozone Gas in Textile Industry

3.3.1 The Use of Ozone Gas in Pre-treatment Process

Due to the many advantages of ozone gas, it is utilized in numerous fields of textile industry. Since ozone gas is an oxidative material, it is an alternative to hypochlorite, chlorite and hydrogen peroxide in bleaching process. Compared with other bleaching agents, it does not generate waste, with less damages to textile fabrics and no harm to environment as well as human health. In textile bleaching and washing, ozone gas provides energy and water saving, reducing the use of washing chemicals and decreasing the duration of process. Furthermore, as a new, dry, inexpensive and eco-friendly surface treatment for solid surfaces, ozone gas combined with UV-radiation is utilized in surface modification of textile fibers. Because of high oxidative properties of ozone gas with UV-radiation, surface adhesion can be improved, resulting in the production of high quality products. Ozone and UV-radiation has also etching effect which influences the surface wetting properties [10].

Recently, researchers have investigated the use of ozone in bleaching process of textile materials because of its high oxidizing capacity and opportunities. Perincek et al. studied the use of ozone gas in bleaching cotton fabrics. According to the results, cotton fabric can be bleached in a short time with ozonation treatment [11].

Kan et al. investigated the effect of plasma-induced ozone treatment on the color fading of reactive dyed cotton fabric. In this context, cotton fabric was dyed with yellow reactive dye and then the dyed samples were treated with plasma-induced ozone under different conditions. The fading behaviors of samples were tested with spectrophotometer. The results showed that the plasma-induced ozone treatment reduced the processing steps and cost in comparison to conventional process [12].

In other study, jean fabric was treated with ozone injected water combined with ultrasound and hydrogen peroxide. A combined process effect on dye degradation was tested with electron paramagnetic resonance spectroscopy. According to the results, ozone was more effective with respect to bleaching of the jean samples than the hydroxyl radicals when combined with ultrasonic energy. The ultrasonic cavitations improved the diffusion of ozone through the fabric, resulting in the degradation of indigo dye. Furthermore, the use of moderate concentration of ozone caused no damage to cotton fibers [13].

Perincek et al. investigated the effects of ozonation on dyeing and bleaching properties of Angora rabbit fibers. The results showed that ozonation causes to improve the degree of whiteness and dyeability of Angora rabbit fibers [14].

In another study, cotton fabric samples were treated with ozone and ultrasound combination instead of conventional methods. After the pre-treatment, samples were dyed with different plant-based natural dyes. The use of ozone and ultrasound combination in pre-treatment caused not to use mordant agents. Furthermore, the fastness properties of dyed samples are good and sufficient for the use [15].

Prabaharan et al. researched the bleaching of grey cotton fabric with ozone/oxygen gas mixture and effects of ozone concentration and treatment time on the properties of bleached fabric. After the bleaching process, whiteness index, strength, elongation, extent of impurities removed, degree of chemical modification and reactive dye uptake were investigated. According to the results, the whiteness index of samples is found to be an acceptable value, with the acceptable results obtained via ozonation in a very short time. Ozone bleaching is ecofriendly since it is not harmful, requiring low quantities of water along with quite short durations of process [16].

3.3.2   The Use of Ozone in Color Remove of Textile Waste Water

In textile industry, among the hardest challenges are the improvements of wastewater after dyeing of textile materials and the amount of water used. In order to produce 1 kg of ready textile, nearly 200 L water is utilized, resulting in a high amount of chemicals. Throughout the world, it is estimated that textile wastewater is loaded with 280,000 t of textile dyes which have aromatic molecular structure. Furthermore, the biodegradation process of these compounds is highly difficult. In addition to these compounds, textile wastewater is charged with other organic and inorganic compounds having toxic effects on the ecosystem. In order to clarify and reutilize textile wastewaters, various methods have been investigated. Physical methods can be listed as filtration, coagulation/flocculation, precipitation, flotation and adsorption. Biological processes are aerobic, anaerobic and combination of them. Physical and biological processes are slower, requiring large storage areas and displaying low efficiencies when it comes to color removal. Chemical processes such as ozonation are generally simpler in application [17].

In the world, textile industry is one of the largest sectors which monopolize 8% of the world trade in manufactured goods. A major problem of textile production is wastewater [18]. The textile industry has adversely affected the environment due to its notorious water consumption and wastewater production. The water demand of the textile industry is estimated as 80–100 m^3 Mg^{-1} [19]. The wastewater of the textile industry includes high amounts of organic and inorganic compounds such as dyes, toxic heavy metals, pentachlorophenol, halogen carries, carcinogenic amines, free formaldehyde, salts and softeners [20]. Owing to all these mentioned facts, the decontamination of textile wastewater becomes crucial. To this end, ozone technology is one of the eco-friendliest techniques among the available techniques.

Textile wastewaters are loaded with different types and concentrations of harmful compounds resulting from textile production steps. Textile wastewaters are indeed charged with highly colored and non-biodegradable colored dyes, surfactants and toxic chemicals. As these wastewaters are very harmful for the ecosystem, new methods have been investigated to meet the quality criteria of water [21]. Biological treatment methods for textile wastewaters are insufficient to meet biological oxygen demand (BOD) and chemical oxygen demand (COD) [22]. Thus, researchers have investigated methods alternative to biological treatment such as adsorption, membrane process and ozonation [22, 23].

Ozone gas is used to remove color from dyed textile wastewaters. During the ozonation process, ozone attacks unsaturated bonds of chromophores which leads to the elimination of color. Furthermore, ozone can degrade complex organic molecules to organic acids, aldehydes and ketones, resulting in the likely removal of color molecules. However, due to some difficulties arising from the high cost of ozone production and low ozone solubility and stability in water, an advanced oxidation process is required. Some significant oxidation processes can be listed as those being used with ozone as well as UV radiation and some chemicals [23].

In literature, the use of ozone gas was investigated with respect to textile wastewaters. Pazdzior et al. investigated the acute toxicity of textile wastewaters before and after chemical and biological treatments separately as well as a combination of chemical-biological treatments. According to the results, biodegradation followed by ozonation led to the highest toxicity reduction [24].

Bilińska et al. studied ozonation of textile wastewater discharged with Reactive Yellow 145, Reactive Red 195 and Reactive Blue 221. In this study, the effects of four ozonation process were investigated such as O_3, O_3/H_2O_2, O_3/UV and $O_3/UV/H_2O_2$. According to the results, it can be concluded that ozonation resulted in fast decolorization followed by further decomposition of by-products [25].

Reactive dyeing of cotton fibers brings about colored wastewater including residual dyes, electrolyte, alkali and other auxiliaries. Hu et al. investigated the reuse of reactive dyeing bath through catalytic ozonation with novel catalysts. In order for degradation, two novel ozonation catalysts, mesoporous carbon aerogel and supported cobalt oxide nanoparticles were produced. Degradation efficiency was obtained with decolorization and chemical oxygen demand. According to the results, novel ozonation catalysts improved the decolorization and oxygen demand removal [26].

Recently, the use of ozonation as a pre-treatment before biological process has been largely studied in relation to the purification of textile wastewaters. Punzi et al. studied the use of an aerobic biofilm reactor followed by ozonation of real textile wastewater including azo dyes. Acute toxicity tests were carried out before and after ozonation. According to the results, the combination of anaerobic-ozonation processes caused to remove more than 99% of color, 85–90% of chemical oxygen demand and toxicity [27].

Cardoso et al. developed a bubbling annular reactor which provides to test the efficiency of photolysis, photocatalysis, photoelectrocatalysis and ozonation using oxygen or ozone gas flow. After bubbling of ozone, the results showed that 90% of color was removed [28].

3.3.3   The Use of Ozone in Sizing

In the textile industry, before the weaving, natural based warp yarns are treated to sizing process to stabilize the high-speeds of weaving. In order to obtain acceptable sizing properties for warp yarns, polyvinyl alcohol (PVA) has been widely used for decades. Though many advantages of PVA sizing, decomposition of PVA wastewater is difficult in nature with the high level chemical oxygen demand (COD). Because of these difficulties, researchers have sought for green sizing recipes and environmental production methods.

Sun et al. used phosphate-modified starch (PM-starch) and glycerol to sizing cotton yarns. According to the results, glycerol caused to decrease sizing pick-up, yarn breaking strength and breaking elongation [29].

Recently, plasma treatment is a clean, dry and green technique to modify surface properties of textile materials.

4   Membrane Filtration Technology

Membrane filtration can be accepted as a very efficient and economical method to separate components dissolved in a liquid. The membrane can be described as a physical barrier which permits compounds to transfer with regards their chemical and physical behaviors. Commonly, membranes include a porous support layer on top of the actual membrane [30].

Membrane filters are selective barriers produced from several materials. Membrane filters are used in various areas in order to separate the compounds smaller than 10 μm from liquor. The flow on the membrane surface occurs in two directions which

Fig. 6 The filtration actions with regard to the pore size [31]

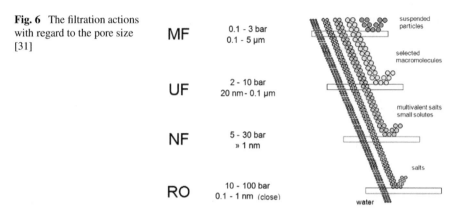

are parallel to the membrane axis and in a radial direction (cross-flow). Particles are bigger than the membrane pores are removed on the surface of the membrane and these particles have parallel flow. Particles are smaller than the membrane pores pass through the membrane with the cross-flow. Finally, the molecules in the liquor are separated physically according to their molecular dimensions [30].

In the industry, microfiltration (MF), ultrafiltration (UF), nanofiltration (NF) and reverse osmosis (RO) are commonly used filtration processes to separate the compounds in the liquor. Figure 6 shows the filtration actions with regard to the pore size [30].

4.1 Types of Membrane Process

4.1.1 Ultrafiltration

Ultrafiltration is a type of filtration method that provides separating extremely small particles and dissolved molecules from fluids. In filtration processes, the most important factor affecting filtration is molecular size. However, in almost all of filtration processes, chemical and physical properties of samples can affect the permeability of filter. In ultrafiltration process, molecules of similar sizes cannot be separated. It means that ultrafiltration can separate molecules having different molecular sizes. Materials having 1–1000 K molecular weights are filtered by ultrafiltration membranes and salt molecules can pass through. Furthermore, materials larger than the membrane pore size pass through the filter, resulting in separation of the contaminants with high molecular weight from the fluid. Sugars, proteins and bacteria can be simply separated with ultrafiltration [30].

Ultrafiltration processes utilized in these fields are as the following;

- Purification of water in laboratory,
- Refine of wastewater,

- Refine of drinking water,
- Recovery of waste dyestuff in the textile and automotive industry,
- Juice and wine industry,
- Pre-treatment for reverse osmosis [32].

Ultrafiltration has been presumed as a practical industrial process due to purification or dewatering of solution. Recently, ultrafiltration process has been used in many industries because of its economic attractiveness and usefulness [33].

4.1.2 Microfiltration

Microfiltration is utilized in order to remove particles having particle size 0.025–10.0 μm from fluids by transferring through a microporous medium. Furthermore, microfiltration process can be used for final filtration or prefiltration [30].

Microfiltration membranes lead all bacteria to move away. However, viruses are not removed with microfiltration membrane as they are of smaller sizes than the pores of a microfiltration membrane. Microfiltration can be implemented in a lot of different treatment processes [30].

Microfiltration process utilized in these fields are as the following;

- Cold sterilization of pharmaceuticals,
- Purification of fruit juice and wines,
- Biological wastewater treatment,
- Separation of oil/water emulsions,
- Pre-treatment of water for nano filtration or Reverse Osmosis,
- Petroleum refining [34].

4.1.3 Reverse Osmosis

Reverse osmosis is described as the finest separation membrane process with pore sizes ranging from 0.0001 to 0.001 μm. Reverse osmosis can remove nearly all molecules without water. Compared with microfiltration, osmotic pressure of reverse osmosis is higher. In reverse osmosis process, osmotic pressure causes to occur chemical potential differences of the solvent [30].

Reverse osmosis filters salts and small molecules from solutions at high pressures using membranes. Reverse osmosis membranes are successfully used to purify water that distilled water quality [30].

Reverse osmosis process utilized in these fields are as the following;

- Removing ions, molecules and larger particles from drinking water,
- Removing bacteria,
- Concentration of fruit juice,
- Syrup production,
- Hydrogen production [30].

4.2 Studies Concerning Membrane Use

Membrane techniques are promising in terms of cleaning textile wastewaters, since dyes stuffs and auxiliary chemicals used for dyeing can be removed with these techniques. Indeed, membrane techniques can be utilized to purify complex wastewaters. In literature, there have been many researches treated to textile wastewaters due to purification. Aouni et al. investigated the reactive dyes' molecular weight and the effects of the membrane types on purification of textile dyeing wastewater. It was obtained that high chemical oxygen demand retention and color retention rates (>90%) were carried out with using ultrafiltration and nanofiltration [35].

In another study, biological treated wastewater was applied to nanofiltration in two ways, direct nanofiltration treatment and nanofiltration after ultrafiltration pretreatment in four different pore sizes. After the treatment, flux, salt retention and COD removal were measured. According to the results, the decrease in pore size of nanofilter caused to improve chemical oxygen demand and salt retention [36].

A membrane bioreactor is defined as combining biological treatment and membrane filtration. Membrane bioreactor systems have been increasingly used as retention of solid particles with membrane ultra-filtration is higher than conventional biological process. Brik et al. investigated the performance of a membrane bioreactor for the textile wastewater originating from a polyester finishing mill. The level of percentage of chemical oxygen demand removal was found as 60–90% and color removal was measured as 87% [37].

Membrane bioreactor treatment to textile wastewaters has been investigated because of simple and significant removal of contaminants and cost-effective process. However, membrane fouling is one of the major drawbacks which leads to decrease in permeate flux. Jegatheesan et al. investigated aerobic and anaerobic membrane bioreactor process for textile wastewater treatment. It has been found that long sludge retention time increases the degradation of pollutants [38].

Lutz et al. performed zwitterionic copolymer membranes for industrial wastewater streams. These membranes were prepared with self-assembling zwitterionic amphiphilic random copolymer on porous supports. According to the results, these membranes are effective to polysaccharides, natural organic matter and fatty acids [39].

In other study, biological treated textile effluent is filtrated with nanofiltration membrane. Then color removal and chemical oxygen demand reduction were analyzed. It was found that hollow nanofiltration membranes caused effective color removal and chemical oxygen demand reduction [40].

Zhu et al. investigated the removal of reactive dye from textile waste water with nanofiltration. The nanomembrane was produced with hollow fiber in laboratory conditions. The efficiency of nanofiltration was evaluated according to dye removal efficiency. According to results, under the pressure of lower than 1 bar, the membrane filtrated all the dye molecules in the textile waste water. The efficiency of membrane increased with increasing in the transmembrane pressure [41].

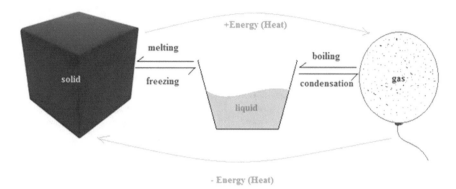

Fig. 7   States of matter

Fig. 8   The fourth state of material: plasma [42]

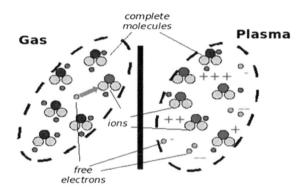

5   Plasma Technology

Plasma can be defined as the fourth state of material having the highest energy. In order to identify plasma, states of matter must be explained. When energy is given to a solid material, the distance between atoms increases, carrying out their vibrations more freely. In the present case, the solid material melts into liquid form. If energy delivery to this material goes on atomic mobility increases, and when the material receives the necessary energy for evaporation atoms begin to move freely in random directions, going from the solid phase to the gaseous state (Fig. 7).

Upon the maintenance of energy delivery, atoms and molecules begin to decompose into charged particles (ions and electrons), with the material going to the plasma phase.

Briefly, plasma described as ionized gas (Fig. 8).

For the first time, plasma was described by Sir William Crookes in 1879 with Crookes tube. The basis of the Crooks tube is cathode ray which was subsequently identified by Sir J. J. Thomson in 1879. In 1928, Irving Langmuir examined the plasma term in exact terms.

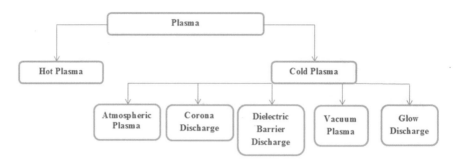

Fig. 9 Types of plasmas

Table 2 The advantages of plasma treatment compared with conventional method

Plasma technology	Conventional methods
– Water is not used in plasma application. Treatments are carried out in gas phase	– Water-based method
– Low water, energy and chemical consumption	– High water, energy and chemical consumption
– Short period of application	– Long period of application
– Plasma application does not affect the bulk properties of materials	– Bulk properties of materials are generally affected
– Complex and multifunctional	– Simpler
– Electric energy is used	– Heat energy is used

In plasma applications, reactive particles emerged, leading to modifications on the surface of materials. These modifications can be classified as surface activation, corrosion, grafting, cleaning and cross-linking.

Basically, plasma treatment can be categorized into two groups such as cold and hot plasma. In the textile industry, cold plasmas can be used because it does not cause damage to textile materials. Hot plasma applications lead to carbonization of textile materials. The types of plasmas are shown in Fig. 9.

Plasma technology is utilized in a wide range of textile industry applications due to its numerous advantages. Table 2 demonstrates the advantages of plasma technology compared with conventional methods in the textile industry.

Pre-treatment and finishing of textile materials by using non-thermal plasma applications have recently become more popular. Because of its numerous advantages compared with conventional process, plasma technologies have been preferred. Plasma surface modification does not require the use of water and chemicals. For this reason, plasma treatments are accepted as economic and ecological processes. Furthermore, plasma treatment led to decrease of contaminations as it does not require using water. In the textile industry, non-thermal plasmas are suitable because most of textile materials are sensitive to heat. This technique causes active functional groups to be formed on textile surfaces. In addition, after the non-thermal plasma treatment, textile surfaces can gain wettability, adhesion of coatings, printability, induced oleophobic properties, changing physical or electrical properties, cleaning or disinfection of fiber surface [43].

5.1 Plasma Treatment to Cotton

Plasma treatment can be utilized to modify different types of textile products. In literature, there are a variety of researches using plasma treatment in order to modify cotton fibers. Plasma treatment provides removing PVA sizing [44], increasing hydrophilicity and wickability [45], gaining hydrophobic properties [46], developing adhesion properties [47] and increasing dyeability [48].

Cai et al. applied air/He and air/O_2/He plasma treatment on cotton fabric to desize of PVA. After the plasma treatment, percent desizing ratio and tensile strength were measured. The results showed that plasma treatment removed some PVA sizing and significantly improved percent desizing ratio by washing. The tensile strength of cotton fabric treated with atmospheric plasma is similar to that of unsized fabrics. Furthermore, air/O_2/He plasma is quite effective to remove PVA sizing [49].

In other study, raw cotton fabric samples were treated with air plasma and argon atmospheric plasma. After the plasma treatment, the hydrophilicity and wickability of samples increased and contact angles notably decreased. Furthermore, morphological changes were observed with scanning electron microscope [50].

Li et al. investigated that plasma surface treatment of cotton fabrics were performed in a hexafluoropropene (C_3F_6) atmosphere under different experimental conditions. X-ray photoelectron spectroscopy (XPS) analysis demonstrated that 50% of fluorine atoms were incorporated in the surface structure of two fibers and –CF, –CF_2, –CF_3 groups occurred on the surface. After the contact angle and wet-out time measurements, the fibers demonstrated high hydrophobic properties [46].

Selli et al. investigated RF SF_6 plasma treatment on cotton fabrics. The plasma treatment provided an efficient implantation of fluorine atoms on the surface of both polymers. After the plasma treatment, the fluorinated layer was observed on the surface of cotton fabrics. The fluorinated layer led to the increase in hydrorepellence behaviors of surface [51].

In another study, the hydrophilic improvement of the grey cotton fabric by low pressure dc glow discharge air plasma was investigated. The plasma treatment was achieved for different exposure times, discharged potentials and pressure levels. Effects of plasma treatment on wettability behaviors were investigated. The surface energy values were estimated using contact angle measurement. Furthermore, degradation and dyeability of the fabrics were determined. According to the results, the surface hydrophilicity and energy were found to increase [52].

Sun and Stylios investigated the effects of low temperature plasma treatment on cotton fabrics. In this context, hexafluroethane (C_2F_6) and oxygen plasma treatment were applied to cotton fabrics separately. After the plasma treatment, samples were investigated with regards to the type of gas. The plasma treatment was determined to cause changes on surface geometrical roughness. Furthermore, both rougher and smoother surfaces can be produced using plasma treatment. The plasma treatment led to chemical modification on surfaces [53].

In another study, cotton fabric samples were treated with radio-frequency plasma (in air) at different power levels and time intervals. After the plasma treatment, moisture content and surface resistivity behaviors were investigated with regards to the power level and treatment time. The surface resistivity was found to be affected by both power level and treatment time [54].

Peng et al. investigated the influence of moisture absorption of cotton fabrics on the effectiveness of atmospheric pressure plasma jet on desizing of polyvinyl alcohol. X-ray photoelectron spectroscopy analysis demonstrated that the plasma treated PVA has higher oxygen concentration than the control. Furthermore, the results indicated that the highest desizing efficiency was obtained [55].

In a different study, air and argon atmospheric plasma treatments were applied to bleached plain cotton fabrics. After the plasma treatment, pilling, thermal resistance, thermal conductivity, water vapour permeability, air permeability and surface morphology were investigated. The results showed that the pilling resistance of cotton fabric samples increased. Moreover, thermal resistance, water vapour permeability and surface friction coefficient increased with the plasma treatment. The SEM images demonstrated that the atmospheric plasma modified the fiber surface [56].

5.2 Plasma Treatment to Wool

Plasma treatment has been used by the industry for treatments of metals and other polymeric materials. In the textile industry, plasma treatment has particularly gained greater prominence.

Wool fabrics were treated with oxygen, carbon tetrafluoride and ammonia low temperature plasma. After the plasma treatment, the samples were dyed with natural dyes extracted from cochineal, Chinese cork tree, madder and gromwell. The results showed that the dyeing rate of plasma-treated wool fabrics increased. Furthermore, plasma-treated wool fabrics dyed with cochineal and Chinese cork tree were brighter compared with untreated wool [57].

Ghoranneviss et al. investigated the effects of plasma treatment on the natural dyeing properties of wool fabrics. Before the dyeing process, wool fabric samples were treated with argon plasma. Madder and weld were used as natural dyes and copper sulfate ($CuSO_4$) as a metal mordant. Furthermore, copper was used as the electrode material in a DC magnetron plasma sputtering device. After these processes, the color strength, fatness and anti-bacterial properties of samples were analyzed. The results showed that the plasma treatment led to the development of color strength and fastness properties of samples. Furthermore, the anti-bacterial efficiency of samples improved [58].

In another study, atmospheric pressure plasma was performed to pure cashmere and wool/cashmere textiles with a dielectric barrier discharge in humid air (air/water vapor mixture). Treatment parameters were adjusted in order to develop the wettability of the fabrics. After the plasma treatment, characterization analysis was conducted out to investigate the wettability, surface morphologies, chemical composition and mechanical properties of the plasma treated samples. The analyses revealed a surface oxidation of the treated fabrics that enhanced their surface wettability. Furthermore, SEM analysis demonstrated that a minor etching effect occurred on the textile surface [59].

Zanini et al. investigated the hydro- and oleo-repellent modifications of pure cashmere and wool/nylon textiles by means of an atmospheric pressure plasma treatment after fluorocarbon resin impregnation. The plasma treatment was performed with a dielectric barrier discharge in humid air (air-water vapour mixture). The finishing process was carried out with a foulard system, with an aqueous dispersion of a commercial fluorocarbon resin. After these treatments, wettability, surface morphologies and chemical composition of the modified textiles were investigated in terms of the plasma treatment. According to the results, the plasma treatment caused to increase the hydro- and the oleo-repellent properties of the modified fabrics [60].

In another study, atmospheric pressure plasma treatments were performed to wool/cashmere (15/85%) textiles with a dielectric barrier discharge in nitrogen. The chemical properties of the plasma treated samples were investigated with FTIR/ATR spectroscopy, X-ray photoelectron microscopy and fatty acid gas chromatographic analysis. Furthermore, the mechanical properties of samples were analyzed with KES-F system. The analyses revealed the surface modification of the treated fabrics, which led to the improvement of wettability of samples [61].

Eren et al. synthesized different conducting polymers by atmospheric pressure plasma and coated these conducting polymers on wool fabrics with atmospheric plasma. Scanning electron microscopy, energy dispersive X-ray spectroscopy, Fourier Transform Infrared Spectroscopy and probe resistance measurements were used to investigate the properties of wool samples. After the plasma treatment, electric conductivity of samples increased [62].

In a further study, wool fibers were treated with plasma and a poly (propylene imine) dendrimer to develop dyeing properties. FESEM, EDX, AFM and FTIR analyses were then performed to investigate the effects of these treatments on the chemical and physical properties of wool fibers. According to the results, the etching effect occurred on the surfaces of wool fibers and the plasma treatment caused to increase the roughness of samples. Furthermore, oxygen plasma and dendrimer treatments improved the dyeability of wool fibers with cochineal natural dye [63].

Jeon et al. treated DBD plasma with oxygen, nitrogen, argon and air on wool fibers. In this study, the plasma treatment time was controlled. The increase in plasma treatment time led to the improvement of the wicking rate of wool samples. Furthermore, the surface morphology was investigated with SEM. According to the SEM results, the damages of surfaces increased with increasing treatment time. The wicking rates of wool fibers increase in the order: oxygen, argon, nitrogen and air plasma. As a result, the oxygen plasma is the most effective to change the wettability of the wool fibers [64].

Shahidi et al. investigated the effects of low temperature plasma on the wool fabric samples under different conditions. The reactive gases such as O_2, N_2 and Ar were treated on wool fabrics. After the plasma treatments, morphology, dyeability, hyrophility and fabric shrinkage properties of samples were analyzed. The results showed that surface topography and chemical composition changed after the plasma treatment. The dyeability analysis results illustrated that O_2 and Ar plasma treatments were more effective in terms of increasing the dye exhaustion of wool samples. Moreover, the samples had more brilliant colors with the plasma treatment. The plasma treatment led to the increase in the hydrophility of samples while developing shrink resistance and anti-felting behavior considerably [65].

In another study, wool fibers were treated with low temperature plasma treatment with different gases, namely oxygen, nitrogen and gas mixture (25% hydrogen/75% nitrogen). The results showed that chemical composition of the plasma treated wool fibers varied differently with different plasma gas [66].

Moreover, wool fabrics were treated with plasma of different gases (air, oxygen, water vapor (H_2O) and nitrogen) for different periods of time. The surface changes of wool samples were analyzed with XPS. According to the results, air plasma treatments oxidized and etched the F-acid mono-layer. Furthermore, after the nitrogen plasma treatment, new nitrogen groups were not observed on the surface. By contrast, oxygen gas led to the oxidation of the surface of wool fibers and their posterior etching [67].

5.3   Plasma Treatment to Synthetic Fibers

Jurak et al. prepared new mixed chitosan/phospholipid films and applied the films on plasma activated PET surfaces. The prepared surfaces were characterized with regards to the wettability and surface thermodynamics. According to the results, it was thought that the PET surfaces could be used in reducing inflammation and accelerating wound healing [68].

In another study, it was aimed to improve the adhesion properties of PET films using atmospheric plasma. PET is known as chemically inert to most coatings, but the surface can be modified with the atmospheric plasma treatment. Furthermore, after the plasma treatment, the surface becomes enriched with oxygen, rougher and more wetting. Lastly, adhesion test demonstrated the improvement of adhesion after the plasma treatment [69].

In another study, the effects of air plasma treatment on PET and starch modified PET were investigated. Air plasma treatment was applied in order to develop the interfacial adhesion of starch to PET. After the plasma treatment, wettability, thermal and mechanical properties of samples were investigated. After the plasma treatment for a short period of time, the contact angle of samples decreased immediately. Furthermore, plasma treatment led to the differences of elongation as well as small differences in thermal stability and flexural properties of PET [70].

Zhang et al. applied low-pressure oxygen and argon plasma to nylon and modified fabrics. After the plasma modification, the fabric samples were coated with single-walled carbon nanotubes by a dip-drying process. After the coating process, sheet resistance, fiber surface roughness and the attachment of single-walled carbon nanotubes onto nylon fabrics were investigated. The plasma treatment caused to increase the roughness of samples and the attachment of single-walled carbon nanotubes onto nylon fabrics. After the coating process, the sheet resistance of samples also improved [71].

In another study, multi-walled carbon nanotubes were modified by combination of oxygen + nitrogen in order to improve its dispersion in the nylon matrix and enhance the interfacial adhesion. After the plasma treatment, the tensile strength, Young's modulus, elongation at break and storage modules were improved by ~66%, 64%, 69% and 39%, respectively. It was thought that the plasma treatment improved interfacial adhesion [72].

Sanaee et al. investigated the effects of oxygen and hydrogen radio frequency plasma on PET films with SEM, XPS and atomic force microscopy (AFM). It was found that the plasma treatment led to the reduction in penetration of air through the PET films. Furthermore, compared with hydrogen plasma, oxygen plasma resulted in a rougher surface [73].

In another study, naylon 6 fabrics were treated with low temperature plasma with three gases: oxygen, argon and tetrafluoromethane. After the plasma treatment, the properties of fabric such as surface morphology, low-stress mechanical properties, air permeability and thermal properties were analyzed. The different plasma gases resulted in different morphological changes. Low-stress mechanical properties were analyzed with Kawabata evaluation system fabric (KES-F). The results showed that surface friction, tensile, shearing, bending and compression properties of fabric samples changed with the plasma treatment. In addition, the plasma treatment caused to decrease the air permeability of samples probably due to plasma action resulting in an increase in fabric thickness and a change in the surface morphology. Finally, thermal properties of samples improved with the treatment [74].

McCord et al. treated the atmospheric pressure He and He–O_2 plasma to polypropylene and nylon 66 fabrics for selected time intervals of exposure. Scanning electron microscope showed no changes in the samples. After the plasma treatment, carbon and oxygen contents of naylon 66 altered. The oxygen and nitrogen contents of polypropylene fabric samples increased significantly. In addition, plasma treatment leads to decrease in tensile strength of naylon 66 fabric samples [75].

In another study, acrylic fabric samples were treated with a RF atmospheric pressure plasma, after which a fluorocarbon finish was applied to the samples. In the plasma application, helium and helium/oxygen gas were used for different periods of time. The oil and water repellency of samples were investigated with regards to the gas type and processing time. After the plasma treatment, the repellency properties of samples improved. Furthermore, plasma treatment caused to change the morphology of samples [76].

Liu et al. aimed to improve the antistatic properties of acrylic fibers by using nitrogen glow-discharge plasma. The treated surfaces are characterized by scanning electron microscopy, specific surface area analysis and X-ray photoelectron spectroscopy. The plasma treatment led to the increase in surface roughness, wettability and antistatic ability [77].

6 Enzymatic Treatments

Enzymes are biocatalysts composed of metabolic products of live organisms acquired from bacterial derivatives. Catalysts are the substances involving in chemical or biochemical reactions and remaining unchanged at the end of the reaction [78].

All of the isolated enzymes have a protein structure or consist of a protein component. They are named by attaching an "ase" affix at the end of the substance that they affect or based on the reaction that they catalyze. The compound that an enzyme affects is called substrate and the number of the substrate molecules that they affect in a second is called the enzyme turnover number. These substances function as a group within a cell and the final product of an enzyme acts as the substrate of the next enzyme; for instance enzyme amylase converts starch to two-chain maltose and enzyme maltase converts one-chain maltose to glucose [78].

Enzymes are used extensively for many years in medicine, analyses, food chemistry, beverage industry and home-type detergents. 80% of detergents used at homes consist of enzymes. Modern gene technology and enzyme technology will consist of novel types with fixed production and potential extensive applications that will continue in the future [78].

Amylases used to decompose starch in textile industry are known since 1910s. Interesting applications were discovered for wool finish during the recent years. The flake layer on the surface is removed by the utilized enzymes in these methods and wool fabric surface is modified. Furthermore, anti-felting property is introduced and shininess and attitude is developed [78].

The oldest patent about these technologies was received 30 years ago. However, this enzymatic cellulose decomposition process did not find an industrial application field in Europe initially. The first successful applications in this field occurred in Japan [78].

The functions of enzymes can be listed as;

- They reduce activation energy of biological reactions.
- They function only in a specific type of reaction.
- They can function in the same type of reaction without decomposing.
- They enable the reactions to reach equilibrium quickly.
- They function in inorganic environments as well.
- They initiate their reactions at the external surface of the substance that they affect.

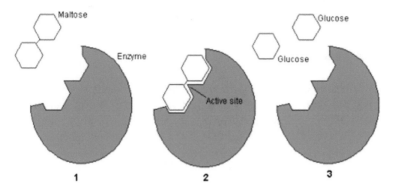

Fig. 10 The key-lock compatibility model [78]

6.1 Enzyme Structure

Genes encode all enzyme proteins. Hence, their amino acid sequence is unique. Some enzymes (e.g. pepsin and urease) are composed of only proteins. However, the other majority is composed of two different parts. These are;

- Protein Component (apo-enzyme part of the enzyme): This part determines which substance will be affected by the enzyme.
- Coenzyme Component: It is organic or inorganic and composed of phosphate in general, and it is an extremely small molecule in comparison to the protein component. Organic molecules required for enzyme action are called "coenzymes". This component is responsible for enzyme action and carries out the main function. Some enzymes are active when specific ions are added in the material [78].

6.2 Enzyme Action Mechanism

Apo-enzyme component determines the type of substrate with which the enzyme will function. There is a relation between the apo-enzyme portion and substrate. German chemist Emil Fischer suggested that this resembles the key-lock fit. Coenzyme component functions closely to the chemical bond generally, for instance it breaks ester bonds. It is thought that apo-enzyme component of enzyme adheres or binds to the substrate molecule (at its active regions) (enzyme-substrate complex) and the coenzyme component combines or binds with the bonds on the substrate in real sense meanwhile and it digests it. A schematic illustration of the key-lock fit is shown in Fig. 10 [78].

6.3 Factors of Affecting Enzyme Action

6.3.1 Temperature

Enzymes are inactive at high temperatures. When the temperature goes up to 100 °C, the reaction rate doubles in other words the reaction rate-increase is proportional to temperature. However, it starts to drop at a specific point and stops completely. The temperature where enzymes act the best is called optimum temperature (generally 55–600 °C) [78].

6.3.2 pH

Enzymes are extremely sensitive to pH changes. They are inactive in extremely acidic and alkaline environments generally. Enzymes show highest activity in a specific pH value in some cases. This pH value is called optimum pH. For example, pepsin, which digests proteins, functions in acidic medium of stomach at pH 2. Various electrical charges are generated on the protein molecule depending on pH and the external surface form (tertiary structure) is generated accordingly and the substrate-enzyme compatibility is enabled. This electrical charge possibly improves enzyme-substrate attraction. Strong acids and bases coagulate enzymes [78].

6.3.3 Enzyme/Substrate Concentrations

If pH and temperature is kept constant, a reaction rate occurs depending on the ratio of enzyme/substrate concentrations. Excessive substrate or enzyme may affect this rate in various ways. Enzyme that is to be added in a medium containing abundant substrate will increase the final product amount [78].

6.3.4 Effect of Chemical Substances and Water

Many chemical substances inactivate enzymes, for example cyanide inhibits cytochrome oxidase enzyme that plays an important role in respiration. Death may occur. Fluoride affects enzyme steps converting glucose to lactic acid. Even the enzyme itself may create a toxic effect, for example if 1 mg of crystal trypsin is infused in mice, death occurs. Some snake, bee and scorpion toxins show enzymatic effect as well and disrupt blood cells or other tissues.

Since the majority of enzymes function in water, water amount is also an effective factor for enzyme function. Enzymes are inactive in medium consisting of less than 15% water generally. This factor is essential for jam and syrup production. Diluted jam, honey or syrups ferment and become sour due to this reason. Moreover, water ratio is desired to be less than 15% during grain purchases [78].

6.4 Enzyme Classification

Each enzyme has a 4-digit number, for example in 3.6.1.3 "ATP phosphohydrolase", the first number indicates its class, the second number indicates its subclass, the third number indicates its group and the fourth number indicates its specific row number. According to this, the enzyme classes are as follows:

1. Oxydoreductases: They catalyze redox reactions.
(a) Dehydrogenases: They catalyze electron gaining reactions.
(b) Oxydases: They catalyze electron losing reactions.
(c) Reductases: They reduce substrates by means of a redactor. For example, acetylaldehyde reductase reduces acetylaldehyde to alcohol.
d)   Transhydrogenases: They move hydrogen from one molecule to another and reduce it.
(e) Hydroxylases: They add one hydroxyl or water molecule to their substrates. For example, phenylalanine hydroxylase adds one hydroxyl group to phenylalanine and converts it to tyrosine.
2. Transferase Enzymes: They catalyze transfer of an atom or atom group from one molecule to another molecule (methyl, carboxyl, glycosil, amino, phosphate groups), except for hydrogen.

Decarboxylases: They catalyze CO_2 release from carboxylic acids.

3. Hydrolase Enzymes: They catalyze molecule breaking by adding a water molecule or by means of a water molecule. They also affect ester, peptide, acid-anhydrite and glycosidic bonds.
(a) Esterases: They break ester bonds (lipase, ribonuclease, phosphatase, pyrophosphotase, glycosidase).
(b) Proteases: They break peptide bonds (proteinase).
4. Liases: They break molecules without releasing water molecule. For example, C–C bond is broken by aldolase and decarboxylase. Similarly, there are enzymes breaking C–O and C–N bonds.
5. Isomerases: They make modifications within molecules and change their sequence in space, for example, rasemase and epimerase.
6. Ligases (Synthesases): They catalyze binding of substrate molecules to each other by spending energy, for example, they catalyze activation of amino acids and fatty acids [79].

6.5 Utilization of Enzymes in Textile

The enzymes and their usage in textile industry were given in Table 3 [80].

Table 3 The enzymes and their usage in textile industry

Enzymes	Textile process
Amylase	Desizing
Cellulose	Enzymatic washing of jeans, bio-process, finishing process, regenerated fabrics
Protease	Treatment of protein based fibers such as wool and silk
Catalase	Hydrogen peroxide removal after bleaching
Lactase	Removal of indigo colorants of jean fabrics
Peroxidase	Oxidation of colorants that are not bond covalently
Lipase	Desizing
Pectinase	For bio-cleaning of raw cotton or flax

6.6 Purposes of Bio-process for Textile Finishing

Textile enzyme manufacturers produced cellulases for improving characteristics of cellulosic fabrics and developed new fabric enzymes to be used in finish processes of cotton, linen, ramie and mixtures of these with synthetic fibers. Bio-finishing processes have an important role for textile finishing today. A cellulose preparation is manufactured by modification of nonpathogenic fungi. Enzymatic processes of cellulosic textiles are called bio-finishing process. The purposes of these processes are as follows;

- Cleaning of fabric surfaces and reducing fuzz.
- Improving some features including flexibility and softness.
- Improving fabric flow.
- Improving hydrophilic characteristic.
- Improving colorant attraction, color efficiency, smoothness and shine.
- Providing comfortable wearing characteristics.
- Successful application in fabrics including cotton, linen and viscose with low basis weight without conformity in their elastic and resistance features.
- For stone washing and old look purposes.
- Unmatched softness when combined with classic softeners.
- Resists are prevented by bio-finishing process and surface fuzz is reduced in preliminary processes of fabrics for sharp contoured prints.
- For removing nep and fibers on the cotton surface.
- For achieving touching, flow and natural softness by improving with comfort.
- For permanently preventing fibrillation and bead formation.
- For improving water absorption in especially towels and bath textiles.
- For creating new and original finish effects.
- For ensuring operability in all ecologic processes.

The effects acquired by bio-finishing are permanent even after multiple washing steps. Textile chemicals remain in the fabric and fabric look is destroyed after each wash in classic methods. This method is an alternative competitor of many textile chemicals with its modern content that can be completely digested biologically and does not create pollution. Enzymes provide water and energy savings and reduce waste amount and reduce environmental pollution in industry [80, 81].

6.7 Cellulase

Cellulases are colloidal proteins with high number of molecules in a metabolite form and with high metabolic rate and are obtained from Aspergillus niger, Trichoderma longibrachiatum, Fusarium solani, and Trichoderma viride. Each enzyme molecule in each unit time causes a change in high molecule number. Industrial cellulases are the complexes of cellulase, cellobiase and related enzyme compounds that are not completely uniform. Their molecular weight is 10,000–4,000,000. Enzymes resemble proteins and have primary, secondary, tertiary and quaternary structures. The properties of these substances can change due to alkali, acids, light, temperature, ionization radiation and biological effect factors. These substances have the ability to break 1,4 β-glucoside bonds of cellulose. An enzyme unit is one-unit of measurement. It is related with a group of 1 micromole reduced in the reference system of cellulose. This is its value at favorable pH and temperature conditions (pH 4.6, 40–550 °C).

A reaction field is undisrupted biological cells in the amorphous field of cellulose. Cellulose is not digested statically but it settles accidentally in general depending on the composition of cellulose complex that is dominant in specific porous places.

The effect mechanism of enzymes functions to generate enzyme-substrate complex by enzyme catalysis for instance. Enzymes consist of actual activity centers in a three-dimensional structural form such as grooves, clefts, spaces and packages. Enzyme function generates substrate complexes, for example cellulose-cellulose complex, and the number of factors is effective on this.

Cellulase multi-enzyme complexes have a structure that is not completely uniform to achieve special effects depending on the production methods, and they are also affected by the production conditions that can generate elective cellulose complex reactions. Moreover, substrate properties are also effective for generating elective conformity for enzymatic degradation due to the non-uniform structure of cellulose. In conclusion, thread type and structure, design of textile substrate, and dissociation effect are influential as well. Thin threads and open material structures, and especially each free, accessible, leaped fiber is fit for degradation.

Bio-reaction occurs due to reduction in activation energy for increasing the reaction rate which is a very simple expression of chemical degradation and corrosion in the enzyme-substrate relation mentioned above. In conclusion, the complex is degraded with the release of the reaction product and enzyme that is ready once again.

The application techniques are effective on enzyme activity and fundament. These are temperature, pH (550 °C, pH 4.6) and the diversity of chemical factors. For example, iron, manganese, magnesium and zinc ions and organic salts are affective on generating or inhibiting enzyme activity and on deactivation and improving their characteristics. Irreversible prevention of enzymes is called enzyme toxin [82].

Cellulase catalyzes cellulose chains and enables hydrolytic dissociation [83]. Cellulase has active locations like all enzymes and is a large protein and it catalyzes chemical reactions with its specifically positioned chemical groups. If temperature and pH are not favorable, these active locations change and enzyme activity drops.

It was seen that when American soldiers were at South Pacific Forests, their outfits were worn out extremely and the reasons for that were investigated. After studying thousands of samples, it was found out that the reason was an organism called *Trichoderma viride* (today it is called *Trichoderma reesei*). This organism plays a very important role for improving cellulose enzyme, moreover, it is known as the predecessor of fungi producing enzymes that are used today.

It is interesting that these researches started with the damage of cotton fabrics and they were conducted to prevent their effects that cause hydrolysis. However, we use them to improve their hydrolysis strength. This transition started during 1960s and the army started to produce food and energy products from solid waste by using cellulase enzyme in 1973. 20-fold powerful/effective organisms were generated by using mutant types of *Trichoderma reesei* (obtained in New Guinea).

Today, cellulase is used most extensively in stone washing process in textile industry. However, our knowledge on cellulase is substantially poor in comparison to other enzyme types because cellulose-cellulase systems consist of soluble enzymes that function on insoluble substrates. These systems are very complex. Due to this complexity, extensive researches were conducted and it was concluded in 1950 that this system was a combination of Cl enzyme (removing cellulose crystals = decrystallizing) and hydrolytic enzymes known as Cx (converting cellulose to sugar). This explanation suggested then was discussed and modified for 40 years and some additions were made.

Enzyme complexes of large molecules are not able to enter interior parts of textile materials at the beginning. These are affective primarily on the surfaces where cellulose chains are broken. This is accidental generally. This situation depends on the origin of cellulase complex and especially on its penetration to thin threads and open structure. Microfibers are loose fibers. They break with the effect of biocatalytic degradation and with the mechanical effect on the surface, reduction in tearing strength, wear resistance and weight loss are carried out by process control. After this operation, fabric becomes loose. Smoother and clear surfaces are obtained since fiber ends are removed [82].

At the point reached based on these, the enzyme structure is explained as follows:

– Endoglucanases: They affect dissolved or undissolved glucose chains randomly.
– Exogluconases (a) break off the glucose unit at the end of cellulose (b) cellobiohydrolases break off the cellobiose units (glucose dimers) at the end of cellulose chain.

– Beta glucosidases release D-glucose from the dimer [84].

These there enzyme classes show a synergistic effect and affect cellulase in a complicated way. In conclusion, decrystallization and hydrolysis of natural cellulose becomes possible.

Fungi and bacteria cause digestion of cellulose in nature. These organisms disrupt crystal structure of cellulose by means of some enzymes. Endoglucanases have random effects on cellulose chain or exocellulases affect the terminal groups in cellulose chain.

Cellulases are water insoluble polymers resembling many other enzymes or they are polymeric substrates longer than the enzyme and these are organized as two characteristic places. A catalytic site is connected by flexible bonds to separate the cellulose bonding site. Collaboration is needed between two functional types of two different cellulases for effective digestion of natural cellulose and two different sides of synergetic effect are shown. The collaboration is between endoglucanase and cellobiohydrolase (endo/exo synergetic effect).

It supports creation of the number of free, sensitive terminals for endoglucanase movement and there is collaboration between different CBH. The reason for this synergetic effect and the effects of individual enzymes on crystalline cellulose are not known precisely. CBHs attack cellulose chain at its unreduced terminals [85].

In a study, the stability of cellulase produced from *T. reesei* and its effects on pretreated cellulosic material were studied and it was seen that enzyme stability was good until 450 °C and there was gradual inactivation in enzyme solution at 55–600 °C. The enzyme was completely inactivated in applications at 700 °C for 1 h. It was seen that maximum enzymatic hydrolysis was achieved at pH 6.5 and 500 °C [86]. The effects of mono-components, cellobiohydrolases produced from purified *Trichoderma reesei* and endoglucanases on cotton fabrics were also studied by analysis of weight loss of fabrics, reduction of sugars, and molecular weights of water-soluble oligosaccharides and their generated cotton powders. It was seen that the amounts of sugars reduced in endoglucanases in treatments with cellulase applied with and without mechanic effect and the amounts of dissolved oligosaccharides were high. It was determined that there was reduction in molecular weight of poplin cotton powders for all cellulase types in enzymatic treatments applied without mechanical effect [87]. It is known that structural characteristics of cellulose were effective on the sensitivity of cellulose for enzymatic hydrolysis. The most important characteristics of cellulose fibers including surface area and crystallinity were studied during hydrolysis period, and enzymatic hydrolysis of cellulose and change of structural parameters were studied [88–90].

Catalytic activity of cellulase produced from pure *Trichoderma reesei* in the solution and adsorbed on cotton fabric and the effect of high temperatures on its bonding ability was investigated [91].

X-ray, CP-MAS NMR and structural and morphologic characteristics of four different cellulosic materials were studied and cellulase produced from *Trichoderma viride* and cellobiase enzyme produced from *Aspergillus niger* was applied to cellulosic material obtained from two sources [92].

Commercial cellulases consist of different cellulase mixtures. The characteristics of fabrics treated with these mixed cellulases were studied. The effects of treatment with cellulose mono-components on the molecular and supra molecular structures of cotton cellulose were studied. Desized, bleached print fabrics that were subject to a basic process were produced from ring and OE thread. These fabrics were subject to an enzymatic process with endoglucanase I and II and cellobiohydrolase I and II produced from *Trichoderma reesei*. The application was carried out by using acetate tampon in stainless steel containers containing steel balls. The effects of the enzymatic process were studied in terms of weight loss, reduced sugar formation, fabric breaking strength and tearing strength, copper number, water absorption, hydrogen bonding means and cellulose micro structure and fiber pore size distribution [93].

An enzyme produced from *Penicillum funiculosum* F4 was applied to cotton fibers of four different origins. Sugar was reduced and hydrolyzed in almost all of them within 6 h. There were differences seen in hydrolysis of cottons of different origins. These differences were attributed to the different structural arrangement of cellulose on secondary cell wall of cotton [94].

Cellulase enzymes produced from *Aspergillus niger* and *Trichoderma viride* microorganisms were immobilized and the obtained enzyme preparation was used for digestion of wastes of CMC, cellobiose and filter papers [95].

7   Conclusion

Since prehistoric times, textile materials have been used to cover oneself and protect oneself from climate conditions. With the industrial revolution, textile process has industrialized and textile industry has became a potential environmental pollutant. After the textile wet process, wastewater containing of higher polyaromatic molecular organic compounds have been generates. The quantitative and qualitative profile of textile wastewater show a change according to raw material, production management, technologies, products being manufactured, type and capacity of treatment systems. In this study, the reduction methods of textile wet process pollution were summarized [96]. Owing to the many advantages of ozone, ozone gas is utilized in numerous fields of textile industry such as pre-treatment process and finishing process. Membrane filtration has been accepted as a very efficient and economical method to clean textile waste water. Plasma treatment is a dry and eco-friendly technology as alternative the traditional wet-chemical process.

References

1. Jegatheesan, V., Pramanik, B. K., Chen, J., Navaratna, D., Chang, C. Y., & Shu, L. (2016). Treatment of textile wastewater with membrane bioreactor: A critical review. *Bioresource Technology, 204,* 202–212.

2. Ali, H. (2010). Biodegradation of synthetic dyes—A review. *Water, Air, and Soil pollution, 213*(1–4), 251–273.
3. Kant, R. (2012). Textile dyeing industry an environmental hazard. *Natural Science, 4*(1), 22–26.
4. Aouni, A., Fersi, C., Cuartas-Uribe, B., Bes-Pía, A., Alcaina-Miranda, M. I., & Dhahbi, M. (2012). Reactive dyes rejection and textile effluent treatment study using ultrafiltration and nanofiltration processes. *Desalination, 297*, 87–96.
5. Holkar, C. R., Jadhav, A. J., Pinjari, D. V., Mahamuni, N. M., & Pandit, A. B. (2016). A critical review on textile wastewater treatments: Possible approaches. *Journal of Environmental Management, 182*, 351–366.
6. Calatayud, A., Ramirez, J. W., Iglesias, D. J., & Barreno, E. (2002). Effects of ozone on photosynthetic CO_2 exchange, chlorophyll a fluorescence and antioxidant systems in lettuce leaves. *Physiologia Plantarum, 116*(3), 308–316.
7. Cardoso, J. C., Bessegato, G. G., & Zanoni, M. V. B. (2016). Efficiency comparison of ozonation, photolysis, photocatalysis and photoelectrocatalysis methods in real textile wastewater decolorization. *Water Research, 98*, 39–46.
8. http://reefkeeping.com/issues/2006-04/rhf/index.php, August 22, 2017.
9. https://en.wikipedia.org/index.php?q=aHR0cHM6Ly9lbi53aWtpcGVkaWEuWEub3JnL3dpa2vQ 29yb25hX2Rpc2NoYXJnZQ, August 22, 2017.
10. Rahmatinejad, J., Khoddami, A., Mazrouei-Sebdani, Z., & Avinc, O. (2016). Polyester hydrophobicity enhancement via UV-Ozone irradiation, chemical pre-treatment and fluorocarbon finishing combination. *Progress in Organic Coatings, 101*, 51–58.
11. Perincek, S. D., Duran, K., Korlu, A. E., & Bahtiyari, İ. M. (2007). An investigation in the use of ozone gas in the bleaching of cotton fabrics. *Ozone Science and Engineering, 29*(5), 325–333.
12. Kan, C. W., Cheung, H. F., & Chan, Q. (2016). A study of plasma-induced ozone treatment on the colour fading of dyed cotton. *Journal of Cleaner Production, 112*, 3514–3524.
13. Hamida, S. B., Srivastava, V., Sillanpää, M., Shestakova, M., Tang, W. Z., & Ladhari, N. (2017). Eco-friendly bleaching of indigo dyed garment by advanced oxidation processes. *Journal of Cleaner Production, 158*, 134–142.
14. Perincek, S., Bahtiyari, M. I., Körlü, A. E., & Duran, K. (2008). Ozone treatment of Angora rabbit fiber. *Journal of Cleaner Production, 16*(17), 1900–1906.
15. Benli, H., & Bahtiyari, Mİ. (2015). Combination of ozone and ultrasound in pretreatment of cotton fabrics prior to natural dyeing. *Journal of Cleaner Production, 89*, 116–124.
16. Prabaharan, M., Nayar, R. C., Kumar, N. S., & Rao, J. V. (2000). A study on the advanced oxidation of a cotton fabric by ozone. *Coloration Technology, 116*(3), 83–86.
17. Cardoso, J. C., Bessegato, G. G., & Zanoni, M. V. B. (2016). Efficiency comparison of ozonation, photolysis, photocatalysis and photoelectrocatalysis methods in real textile wastewater decolorization. *Water Research, 98*, 39–46.
18. Punzi, M., Nilsson, F., Anbalagan, A., Svensson, B. M., Jönsson, K., Mattiasson, B., et al. (2015). Combined anaerobic–ozonation process for treatment of textile wastewater: Removal of acute toxicity and mutagenicity. *Journal of Hazardous Materials, 292*, 52–60.
19. Rosi, O. L., Casarci, M., Mattioli, D., & De Florio, L. (2007). Best available technique for water reuse in textile SMEs (BATTLE LIFE Project). *Desalination, 206*(1–3), 614–619.
20. Jadhav, S. B., Chougule, A. S., Shah, D. P., Pereira, C. S., & Jadhav, J. P. (2015). Application of response surface methodology for the optimization of textile effluent biodecolorization and its toxicity perspectives using plant toxicity, plasmid nicking assays. *Clean Technologies and Environmental Policy, 17*(3), 709–720.
21. Polat, D., Balcı, İ., & Özbelge, T. A. (2015). Catalytic ozonation of an industrial textile wastewater in a heterogeneous continuous reactor. *Journal of Environmental Chemical Engineering, 3*(3), 1860–1871.
22. Liakou, S., Pavlou, S., & Lyberatos, G. (1997). Ozonation of azo dyes. *Water Science and Technology, 35*(4), 279–286.
23. Polat, D., Balcı, İ., & Özbelge, T. A. (2015). Catalytic ozonation of an industrial textile wastewater in a heterogeneous continuous reactor. *Journal of Environmental Chemical Engineering, 3*(3), 1860–1871.

24. Paździor, K., Wrębiak, J., Klepacz-Smółka, A., Gmurek, M., Bilińska, L., Kos, L., et al. (2017). Influence of ozonation and biodegradation on toxicity of industrial textile wastewater. *Journal of Environmental Management, 195,* 166–173.
25. Bilińska, L., Gmurek, M., & Ledakowicz, S. (2017). Textile wastewater treatment by AOPs for brine reuse. *Process Safety and Environmental Protection, 109,* 420–428.
26. Hu, E., Shang, S., Tao, X. M., Jiang, S., & Chiu, K. L. (2016). Regeneration and reuse of highly polluting textile dyeing effluents through catalytic ozonation with carbon aerogel catalysts. *Journal of Cleaner Production, 137,* 1055–1065.
27. Punzi, M., Nilsson, F., Anbalagan, A., Svensson, B. M., Jönsson, K., Mattiasson, B., et al. (2015). Combined anaerobic–ozonation process for treatment of textile wastewater: Removal of acute toxicity and mutagenicity. *Journal of Hazardous Materials, 292,* 52–60.
28. Cardoso, J. C., Bessegato, G. G., & Zanoni, M. V. B. (2016). Efficiency comparison of ozonation, photolysis, photocatalysis and photoelectrocatalysis methods in real textile wastewater decolorization. *Water Research, 98,* 39–46.
29. Sun, S., Yu, H., Williams, T., Hicks, R. F., & Qiu, Y. (2013). Eco-friendly sizing technology of cotton yarns with He/O_2 atmospheric pressure plasma treatment and green sizing recipes. *Textile Research Journal, 83*(20), 2177–2190.
30. Munir, A. (2006). *Dead end membrane filtration.* http://www.egr.msu.edu/~hashsham/courses/ene806/docs/Membrane%20Filtration.pdf
31. https://emis.vito.be/en/techniekfiche/microfiltration, August 22, 2017.
32. http://www.akkim.com.tr/tr/urunler/ultrafiltrasyon/ultrafiltrasyon/i-1621, August 25, 2017
33. Lin, S. H., & Lan, W. J. (1995). Ultrafiltration recovery of polyvinyl alcohol from desizing wastewater. *Journal of Environmental Science & Health Part A, 30*(7), 1377–1386.
34. http://www.lenntech.com/microfiltration-and-ultrafiltration.htm.26.08.2017.
35. Aouni, A., Fersi, C., Cuartas-Uribe, B., Bes-Pía, A., Alcaina-Miranda, M. I., & Dhahbi, M. (2012). Reactive dyes rejection and textile effluent treatment study using ultrafiltration and nanofiltration processes. *Desalination, 297,* 87–96.
36. Gozálvez-Zafrilla, J. M., Sanz-Escribano, D., Lora-García, J., & Hidalgo, M. L. (2008). Nanofiltration of secondary effluent for wastewater reuse in the textile industry. *Desalination, 222*(1–3), 272–279.
37. Brik, M., Schoeberl, P., Chamam, B., Braun, R., & Fuchs, W. (2006). Advanced treatment of textile wastewater towards reuse using a membrane bioreactor. *Process Biochemistry, 41*(8), 1751–1757.
38. Jegatheesan, V., Pramanik, B. K., Chen, J., Navaratna, D., Chang, C. Y., & Shu, L. (2016). Treatment of textile wastewater with membrane bioreactor: A critical review. *Bioresource Technology, 204,* 202–212.
39. Lutz, P. B., Zaf, R. D., Emecen, P. Z. C., &Asatekin, A. (2017). Extremely fouling resistant zwitterionic copolymer membranes with ~1 nm pore size for treating municipal, oily and textile wastewater streams. *Journal of membrane science,* (In press).
40. Zheng, Y., Yu, S., Shuai, S., Zhou, Q., Cheng, Q., Liu, M., et al. (2013). Color removal and COD reduction of biologically treated textile effluent through submerged filtration using hollow fiber nanofiltration membrane. *Desalination, 314,* 89–95.
41. Zhu, X., Zheng, Y., Chen, Z., Chen, Q., Gao, B., & Yu, S. (2013). Removal of reactive dye from textile effluent through submerged filtration using hollow fiber composite nanofiltlration membrane. *Desalination and Water Treatment, 51*(31–33), 6101–6109.
42. Morshed, A. M. A. (2011). An overview of the techniques of plasma application in textile processing. *Textile Today.* http://www.textiletoday.com.bd/an-overview-of-the-techniques-of-plasma-application-in-textile-processing/.
43. Morent, R., De Geyter, N., Verschuren, J., De Clerck, K., Kiekens, P., & Leys, C. (2008). Non-thermal plasma treatment of textiles. *Surface & Coatings Technology, 202*(14), 3427–3449.
44. Cai, Z., Qiu, Y., Zhang, C., Hwang, Y. J., & Mccord, M. (2003). Effect of atmospheric plasma treatment on desizing of PVA on cotton. *Textile Research Journal, 73*(8), 670–674.
45. Karahan, H. A., & Özdoğan, E. (2008). Improvements of surface functionality of cotton fibers by atmospheric plasma treatment. *Fibers and polymers, 9*(1), 21–26.

46. Li, S., & Jinjin, D. (2007). Improvement of hydrophobic properties of silk and cotton by hexafluoropropene plasma treatment. *Applied Surface Science, 253*(11), 5051–5055.
47. Gorjanc, M., Bukošek, V., Gorenšek, M., & Vesel, A. (2010). The influence of water vapor plasma treatment on specific properties of bleached and mercerized cotton fabric. *Textile Research Journal, 80*(6), 557–567.
48. Temmerman, E., & Leys, C. (2005). Surface modification of cotton yarn with a DC glow discharge in ambient air. *Surface & Coatings Technology, 200*(1), 686–689.
49. Cai, Z., Qiu, Y., Zhang, C., Hwang, Y. J., & Mccord, M. (2003). Effect of atmospheric plasma treatment on desizing of PVA on cotton. *Textile Research Journal, 73*(8), 670–674.
50. Karahan, H. A., & Özdoğan, E. (2008). Improvements of surface functionality of cotton fibers by atmospheric plasma treatment. *Fibers and polymers, 9*(1), 21–26.
51. Selli, E., Mazzone, G., Oliva, C., Martini, F., Riccardi, C., Barni, R., et al. (2001). Characterisation of poly (ethylene terephthalate) and cotton fibres after cold SF6 plasma treatment. *Journal of Materials Chemistry, 11*(8), 1985–1991.
52. Pandiyaraj, K. N., & Selvarajan, V. (2008). Non-thermal plasma treatment for hydrophilicity improvement of grey cotton fabrics. *Journal of Materials Processing Technology, 199*(1), 130–139.
53. Sun, D., & Stylios, G. K. (2006). Fabric surface properties affected by low temperature plasma treatment. *Journal of Materials Processing Technology, 173*(2), 172–177.
54. Bhat, N. V., & Benjamin, Y. N. (1999). Surface resistivity behavior of plasma treated and plasma grafted cotton and polyester fabrics. *Textile Research Journal, 69*(1), 38–42.
55. Peng, S., Liu, X., Sun, J., Gao, Z., Yao, L., & Qiu, Y. (2010). Influence of absorbed moisture on desizing of poly (vinyl alcohol) on cotton fabrics during atmospheric pressure plasma jet treatment. *Applied Surface Science, 256*(13), 4103–4108.
56. Karahan, H. A., Özdoğan, E., Demir, A., Aydin, H., & Seventekin, N. (2009). Effects of atmospheric pressure plasma treatments on certain properties of cotton fabrics. *Fibres & Textiles in Eastern Europe, 2* (73), 19–22.
57. Wakida, T., Cho, S., Choi, S., Tokino, S., & Lee, M. (1998). Effect of low temperature plasma treatment on color of wool and nylon 6 fabrics dyed with natural dyes. *Textile Research Journal, 68*(11), 848–853.
58. Ghoranneviss, M., Shahidi, S., Anvari, A., Motaghi, Z., Wiener, J., & Šlamborová, I. (2011). Influence of plasma sputtering treatment on natural dyeing and antibacterial activity of wool fabrics. *Progress in Organic Coatings, 70*(4), 388–393.
59. Zanini, S., Grimoldi, E., Citterio, A., & Riccardi, C. (2015). Characterization of atmospheric pressure plasma treated pure cashmere and wool/cashmere textiles: Treatment in air/water vapor mixture. *Applied Surface Science, 349,* 235–240.
60. Zanini, S., Freti, S., Citterio, A., & Riccardi, C. (2016). Characterization of hydro-and oleo-repellent pure cashmere and wool/nylon textiles obtained by atmospheric pressure plasma pre-treatment and coating with a fluorocarbon resin. *Surface & Coatings Technology, 292,* 155–160.
61. Zanini, S., Grimoldi, E., Citterio, A., & Riccardi, C. (2015). Characterization of atmospheric pressure plasma treated pure cashmere and wool/cashmere textiles: Treatment in air/water vapor mixture. *Applied Surface Science, 349,* 235–240.
62. Eren, E., Oksuz, L., Komur, A. I., Bozduman, F., Maslakci, N. N., & Oksuz, A. U. (2015). Atmospheric pressure plasma treatment of wool fabric structures. *Journal of Electrostatics, 77,* 69–75.
63. Sajed, T., Haji, A., Mehrizi, M. K., & Boroumand, M. N. (2017). Modification of wool protein fiber with plasma and dendrimer: Effects on dyeing with cochineal. *International Journal of Biological Macromolecules*.
64. Jeon, S. H., Hwang, K. H., Lee, J. S., Boo, J. H., & Yun, S. H. (2015). Plasma treatments of wool fiber surface for microfluidic applications. *Materials Research Bulletin, 69,* 65–70.
65. Shahidi, S., Rashidi, A., Ghoranneviss, M., Anvari, A., & Wiener, J. (2010). Plasma effects on anti-felting properties of wool fabrics. *Surface & Coatings Technology, 205,* S349–S354.

66. Kan, C. W., & Yuen, C. W. M. (2006). Surface characterisation of low temperature plasma-treated wool fibre. *Journal of Materials Processing Technology, 178*(1), 52–60.
67. Molina, R., Espinos, J. P., Yubero, F., Erra, P., & Gonzalez-Elipe, A. R. (2005). XPS analysis of down stream plasma treated wool: Influence of the nature of the gas on the surface modification of wool. *Applied Surface Science, 252*(5), 1417–1429.
68. Jurak, M., Wiącek, A. E., Mroczka, R., & Łopucki, R. (2017). *Chitosan/phospholipid coated polyethylene terephthalate (PET) polymer surfaces activated by air plasma.* Colloids and Surfaces A: Physicochemical and Engineering Aspects.
69. Rezaei, F., Dickey, M. D., Bourham, M., & Hauser, P. J. (2017). Surface modification of PET film via a large area atmospheric pressure plasma: An optical analysis of the plasma and surface characterization of the polymer film. *Surface & Coatings Technology, 309,* 371–381.
70. Wiącek, A. E., Jurak, M., Gozdecka, A., & Worzakowska, M. (2017). *Interfacial properties of PET and PET/starch polymers developed by air plasma processing.* Colloids and Surfaces A: Physicochemical and Engineering Aspects.
71. Zhang, W., Johnson, L., Silva, S. R. P., & Lei, M. K. (2012). The effect of plasma modification on the sheet resistance of nylon fabrics coated with carbon nanotubes. *Applied Surface Science, 258*(20), 8209–8213.
72. Roy, S., Das, T., Zhang, L., Li, Y., Ming, Y., Ting, S., et al. (2015). Triggering compatibility and dispersion by selective plasma functionalized carbon nanotubes to fabricate tough and enhanced Nylon 12 composites. *Polymer, 58,* 153–161.
73. Sanaee, Z., Mohajerzadeh, S., Zand, K., & Gard, F. S. (2010). Improved impermeability of PET substrates using oxygen and hydrogen plasma. *Vacuum, 85*(2), 290–296.
74. Yip, J., Chan, K., Sin, K. M., & Lau, K. S. (2002). Low temperature plasma-treated nylon fabrics. *Journal of Materials Processing Technology, 123*(1), 5–12.
75. McCord, M. G., Hwang, Y. J., Hauser, P. J., Qiu, Y., Cuomo, J. J., Hankins, O., et al. (2002). Modifying nylon and polypropylene fabrics with atmospheric pressure plasmas. *Textile Research Journal, 72*(6), 491–498.
76. Ceria, A., & Hauser, P. J. (2010). Atmospheric plasma treatment to improve durability of a water and oil repellent finishing for acrylic fabrics. *Surface & Coatings Technology, 204*(9), 1535–1541.
77. Liu, Y. C., Xiong, Y., & Lu, D. N. (2006). Surface characteristics and antistatic mechanism of plasma-treated acrylic fibers. *Applied Surface Science, 252*(8), 2960–2966.
78. Paulo, A. C., & Gubitz, G. M. (2003). *Textile processing with enzymes* (p. 1855736101). ISBN: Woodhead Publishing.
79. www.genetikbilimi.com; May 05, 2005.
80. Galante, Y. M., & Formantici, C. (2003). Enzyme applications in detergency and in manufacturing industries. *Organic Chemistry, 7,* 1399–1422.
81. Stöhr, R. (1995). Enzymes –biocatalysts in textile finishing, 4, 261–264.
82. Hemmpel, W. H. (1991). The surface modification of woven and knitted cellulose fibre fabrics by enzymatic degradation. *International Textile Bulletin-Dyeing/Printing/Finishing, 37*(3), 5–6.
83. Nicolai, M., Nechwatal, A., & Mieck, K. P. (1999). Biofinish-prozesse in der textilveredlung: moglichkeiten und grenzen. *Textilveredlung, 34*(5–6), 19–22.
84. Rössner, U. (1995). *Enzyme in der baumwoll-vorbehandlung, 30*(3/4), 82–89.
85. Nutt, A., Sild, V., Pettersson, G., & Johansson, G. (1998). Progress curves. *The FEBS Journal, 258*(1), 200–206.
86. Nagieb, Z. A., Ghazi, I. M., & Kassim, E. A. (1985). Studies on cellulase from *Trichoderma reesei* and its effect on pretreated cellulosic materials. *Journal of Applied Polymer Science, 30*(12), 4653–4658.
87. Heikinheimo, L., Miettinen-Oinonen, A., Cavaco-Paulo, A., & Buchert, J. (2003). Effect of purified *Trichoderma reesei* cellulases on formation of cotton powder from cotton fabric. *Journal of Applied Polymer Science, 90*(7), 1917–1922.
88. Fan, L. T., Lee, Y. H., & Beardmore, D. H. (1980). Mechanism of the enzymatic hydrolysis of cellulose: Effects of major structural features of cellulose on enzymatic hydrolysis. *Biotechnology and Bioengineering, 22*(1), 177–199.

89. Paralikar, K. M., & Bhatawdekar, S. P. (1984). Hydrolysis of cotton fibers by cellulase enzyme. *Journal of Applied Polymer Science, 29*(8), 2573–2580.
90. Focher, B., Marzetti, A., Sarto, V., Beltrame, P. L., & Carniti, P. (1984). Cellulosic materials: Structure and enzymatic hydrolysis relationships. *Journal of Applied Polymer Science, 29*(11), 3329–3338.
91. Andreaus, J., Azevedo, H., & Cavaco-Paulo, A. (1999). Effects of temperature on the cellulose binding ability of cellulase enzymes. *Journal of Molecular Catalysis. B, Enzymatic, 7*(1), 233–239.
92. Focher, B., Marzetti, A., Sarto, V., Beltrame, P. L., & Carniti, P. (1984). Cellulosic materials: Structure and enzymatic hydrolysis relationships. *Journal of Applied Polymer Science, 29*(11), 3329–3338.
93. Rousselle, M. A., Bertoniere, N. R., Howley, P. S., & Goynes, W. R., Jr. (2002). Effect of whole cellulase on the supramolecular structure of cotton cellulose. *Textile Research Journal, 72*(11), 963–972.
94. Bhatawdekar, S. P., Sreenivasan, S., Balasubramanya, R. H., & Sundaram, V. (1992). Enhanced enzymolysis of never-dried cotton fibers belonging to different species. *Journal of Applied Polymer Science, 44*(2), 243–248.
95. Simionescu, C. I., Popa, V. I., Popa, M., & Maxim, S. (1990). On the possibilities of immobilization and utilization of some cellulase enzymes. *Journal of Applied Polymer Science, 39*(9), 1837–1846.
96. Nadeem, K., Guyer, G. T., & Dizge, N. (2017). Polishing of biologically treated textile wastewater through AOPs and recycling for wet processing. *Journal of Water Process Engineering, 20*, 29–39.

Chapter 3
Sustainability in Wastewater Treatment in Textiles Sector

P. Senthil Kumar and A. Saravanan

Abstract Wastewater is a noteworthy natural hindrance for the development of the textile industry other than the significant issues like ecological contamination. Wastewater treatment, recycle, and reuse have now turned out to be critical interchange wellsprings of water supply. Wastewater is utilized water from local, business, mechanical, and farming exercises. In this chapter, distinctive treatment techniques to treat the wastewater have been discussed. Treating wastewater requires a thorough arranging, plan, development, and administration of treatment offices to guarantee that the treated water is all right for human utilization and for release to the earth. The potential treatments incorporate primary, secondary, and tertiary treatment utilizing physical, chemical and biological processes. The economic pointers chose were assets, process and administration, and client expenditure since they decide the financial moderateness of a specific innovation to a group. Ecological markers incorporate vitality utilize, in light of the fact that it in a roundabout way measures asset usage and execution of the innovation in expelling traditional wastewater constituents. Low-cost by-products from agricultural, industrial and household parts has been perceived as a reasonable answer for wastewater treatment. They permit accomplishing the expulsion of poisons from wastewater and at same time to add to the waste minimization, recuperation and reuse.

Keywords Wastewater treatment · Sustainable technologies · Green built environment · Biological treatment

P. Senthil Kumar (✉)
Department of Chemical Engineering, SSN College of Engineering, Chennai 603110, India
e-mail: senthilchem8582@gmail.com

A. Saravanan
Department of Biotechnology, Vel Tech High Tech Dr. Rangarajan Dr. Sakunthala Engineering College, Chennai 600062, India

© Springer Nature Singapore Pte Ltd. 2018   67
S. S. Muthu (ed.), *Sustainable Innovations in Textile Chemical Processes*,
Textile Science and Clothing Technology, https://doi.org/10.1007/978-981-10-8491-1_3

1 Introduction

Now that sustainability is getting to be a major concern, it is essential that water issues are comprehended in a more incorporated and creative way. With a specific end goal to create maintainable wastewater treatment it is expect to see the wastewater treatment frameworks utilizing an all-encompassing methodology [1, 2]. An all-encompassing methodology infers considering the essential and auxiliary natural impacts and costs that the frameworks create. Chemicals are the contamination delivered at the power plant (creating power for wastewater treatment) and the vitality cost of delivering treatment chemicals. Outlining or choosing a treatment framework in light of maintainability criteria includes a multidisciplinary approach where engineers collaborate with social researchers, financial analysts, researcher, wellbeing authorities and the general population.

In this regard, wastewater can see as a bearer of assets and vitality. After supplements and natural vitality have been recuperated, emanating, as a side effect, would then be able to be reuse. This is unique to customary procedures, as it seeks after gushing as a fundamental item paying little respect to different assets and vitality recuperation. Natural vitality recuperation can contribute extensively to lessening waste generation and CO_2 discharges, and phosphate recuperation can mitigate consumption of phosphorus stores on Earth. Additionally, using or recouping build-ups from drinking water, reaping storm water and notwithstanding delivering biofuel with wastewater and microalgae would all be able to add to supportability in water utilize [3, 4].

Wastewater accumulation frameworks (drain systems) and concentrated and cluster treatment frameworks are compose furthermore, oversaw essentially to ensure human and ecological wellbeing. Despite the fact that their advantages are generally perceive, there are different parts of this foundation furthermore, related advancements that are undeniable and henceforth less recognized, yet they affect groups and the encompassing condition. For example a affirmative part of the open deplete sort out is the gathering and move of wastewater to fitting management workplaces, whereby pathogens and engineered constituents, for instance, oxygen draining regular issue and phosphorus are emptied earlier than the indulgence water is returned to nature [5]. A harmful part of such a system is, to the point that it can make lopsidedness in water, supplement motions, and in this manner contort regular hydrological and natural administrations. For example, the release of vast quantity of indulgence waste water that hold little convergences of concoction component might still prompt an extreme contribution of supplements in a getting water body, in this way, prompting a water quality issue. While there is no agreement on the meaning of maintainability, what is clear is that it makes progress toward the upkeep of monetary prosperity, insurance of the condition and judicious utilization of normal assets, and impartial social advance which perceives the simply needs of all people, groups, and nature. Besides, it perceives the need to outline human and modern frameworks that guarantee humanity's utilization of common assets and cycles do not prompt reduced nature of existence owing to any misfortunes in future monetary open doors or unfriendly effects on community circumstances, human wellbeing and the earth [6].

The Sustainability operation of wastewater treatment frameworks can be surveyed throughout various evaluation devices such exergy investigation, financial examination, and existence phase evaluation. The utilization of an adjusted arrangement of pointers that gives a comprehensive evaluation was picking for assessing the supportability of the diverse wastewater treatment advances. These wastewater treatment innovations incorporate mechanical frameworks, tidal ponds frameworks, and arrive treatment frameworks. Mechanical frameworks, for example, initiated muck use physical, synthetic and natural instruments to expel supplements, pathogens, metals and other harmful mixes [7, 8]. Tidal pond frameworks utilize physical and organic procedures to indulgence waste water, though arrive management frameworks use soil and vegetation, devoid of huge requirement for reactors and active work, vitality and element.

Sustainability is mainly characterized as the fitting incorporation of natural fineness, financial fortune and social impartiality. In fact, the thought of manageability stresses the indivisible fuse of economy, condition and welfare. Different investigations contend that it is an especially imperative errand to characterize and translate the pith of maintainability prior to any green outline executions [9]. Hence, broad surveys relating to the ramifications of manageability (with a outlook to its excellence pointers) has been created.

A supportable building is portrayed by the accompanying basics:

- Demand for safe building, adaptability, advertise and monetary esteem
- Neutralization of natural effects by including its unique circumstance and its recovery
- Human prosperity, tenants fulfilment and partners rights
- Social equity, tasteful enhancements and safeguarding of social esteems.

Wastewater reuse is winding up especially vital in that region; the water asset is instinctively rare. Then again, the Water system mandate likewise gives the bases to accomplish a feasible utilization of water as long as possible, considering ecological, financial and social contemplations. Choosing a feasible treatment for wastewater reuse offices exhibit a genuine test for venture chiefs and in addition for different partners and performing artists associated with the basic leadership process. Generally, the fitting behaviour innovation can be figured out by some factors. These components rely upon the uncommon desires and qualities of all sites, so they might be not quite the same as place to put. Along these lines, because of logical contrasts that exist among nations, a fitting innovation for one particular site will not be appropriate for another. The intricacy of the issue comes from the presence of various components that affect the choice of fitting innovation [10]. In this way, for all casing, few mixes of waste water recovery also, reclaim medications. Also, the proper sterilization innovation, or reuse level ought to be contemplated particularly for every treatment plant. Consequently, substantial and immaterial criteria should be examined together. This inquire about work has concentrated on executing supportability criteria in basic leadership for choosing wastewater sterilization innovation.

Table 1 Sustainable technologies of wastewater generation

S. no.	Regeneration process	Type	Technology
1	Pre-treatment	(i) Physical (ii) Physical and chemical (iii) Physical and biological	(i) Sand filter, ultrafiltration (ii) Coagulation and flocculation and sedimentation/filtration (iii) Infiltration-percolation, constructed wetlands
2	Disinfection	(i) Physical (ii) Chemical (iii) Biological	(i) Ultraviolet radiation, reverse osmosis (ii) Chlorination, ozonation (iii) Regular Systems (development lakes, built wetlands)

Every recovery innovation have firm attributes and the choice of appropriate development for each recover undertaking should be finished depending different parts, as well as the value and quantity of water to recover, the value that should be gone after utilize, the assets expenses, the process and upkeep expenses, arrive prerequisites, the unwavering quality, and natural and communal criterion. Accordingly, the reasonable system is what gives superior execution at a lesser price, yet not just, should likewise consider what is maintainable as far as addressing nearby needs. In this way, the issue is to choose the ideal accessible innovation to actualize in a specific site, and to gather a particular waste water treatment purpose. The recovery medicines can be grouped into pre-treatments and cleaning medications. The pre-treatments are an earlier advance to purification and intend to set up the water for legitimate cleaning, evacuating solids and natural issue, for the most part. Sanitizations medicines decrease intensity of pathogen, and might likewise signify including a lingering point of decontaminator to recycled water. Nevertheless, one ought to keep away from the age of disinfectant. A portion of the fundamental existing advances of wastewater recovery is record in Table 1. Then again, the advancement of new advances and the change of a portion of the current innovations, making them financially aggressive, have made significant troubles in choosing an ideal innovation for a particular case.

2 Resources in Domestic Wastewater and Organic Household Waste

Considerable measures of plant supplements and natural issue are available in family waste and waste from nourishment preparing enterprises. Hypothetically, the supplements in household wastewater and natural waste are about adequate to prepare products to encourage the total populace. This, nevertheless, requires that individuals swing to a vegan slim down. It additionally requires that suitable advances are accessible for safe reusing of the wastewater assets. Essentially 20–40% of the water utilization in sewered urban communities is utilized to flush toilets. With a specific end goal to develop towards a feasible society, require reusing supplements, lessening the water utilization, and limit the vitality required to work squander treatment forms.

While reusing local natural waste cannot supplant mineral manure altogether, it can diminish contamination from household squander, decrease unnecessary manure utilize and create soils that are more advantageous. Tertiary treatment offices can be intend to expel both nitrogen and phosphorous, yet, reusing of the nitrogen is troublesome unless nitrogen is encouraged as struvite or evacuated utilizing smelling salts stripping with adsorption. The most widely recognized strategy for nitrogen expulsion in regular treatment plants today are natural procedures. Nevertheless, with these techniques the vast majority of the expelled nitrogen is release to climate. Phosphorus is most usually evacuated by concoction precipitation utilizing either Fe-or Al-salts as accelerating operators. Nonetheless, the plant accessibility of phosphorus hastened as Fe-or Al-phosphates can be extremely restricted because of low solvency under ordinary soil conditions though with lime precipitation the phosphates are less demanding broken up and accessible to the plants. Since enterprises, families, and road overflow release to a similar sewer framework, there is a danger of substantial metals and different contaminants [11].

2.1 Wastewater Types, Sources and Constituents

2.1.1 Residential Wastewater Framework

Recognizing the nature (sorts, sources and constituents) is basic for outline and choice of wastewater treatment advancements for various areas. The Starting point and streams of wastewater in an urban situation was shown in Fig. 1.

The majority of the wastewater is created from is classification and primarily from families. Despite the fact that the amount and quality is dependent on the accessible water provided, populace measure, way of life and climatic conditions and so forth. Largely, family units create an expected 80 for each penny of the aggregate waste water; both dark waste water and dim waste water (from lavatory and clothing). Human body squanders (dung and urine) delivered from fam-

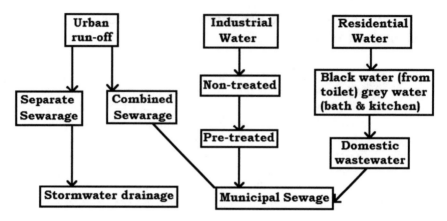

Fig. 1 Starting point and streams of wastewater in an urban situation

ilies or business offices, comprises overwhelmingly of solids, that is the natural bit (starches and fats), and the fluid element fundamentally of coarseness, metal and salts of the waste water. For example, 135–270 g of wet solids (excrement or natural bit) and a comparing 1.0–1.3 kg of pee (fluid part) are delivering per individual every day in creating nations. Families with low water utilization of 40–100 l/individual/day delivers around 70 for each penny sewage with solid BOD of the range $BOD_5 = 300–700$ mg/L, would entail supplementary oxygen to oxidize the wastewater. It ought to be renowned, both the excrement and pee compose a great many intestinal microscopic organisms and predetermined quantity of creatures, the larger part of which are safe and some are gainful, while others are malady causing to people.

This is a wastewater framework that procedures wastewater from a home, or gathering of homes. The framework incorporates the source of wastewater in the home, advances for treating the wastewater, and advances and procedures for returning the handled wastewater to the biological system. Figure 2 shows that disentangled representation of this aggregate wastewater framework for a single home. It involves:

- The home itself–how it is manufactured may influence how wastewater is made
- The advances in the home, for example, clothes washers and toilets
- Sustenance (supplements), family unit cleaners and water
- The general population and their conduct
- The subsequent wastewater
- Reusing and treatment–on location or off-site
- The biological system inside which the house is insert.

Fig. 2 Domestic wastewater
system

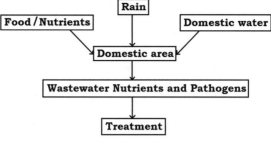

Fig. 3 Industrial wastewater
system

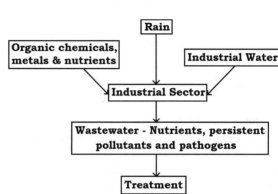

2.1.2   Industrial Wastewater Framework

Industrial sectors and procedures constitute assortments of determined and need contaminations show in wastewater, break up inert mixes, hydrocarbons, oil, and salts and so forth. In spite of the fact that they occur in various stages of poisonous quality and their trademark fluctuates by modern entities, their far reaching and defilement have not been widely identified. Sewage created from raw petroleum refinery, deliver substantial assortments of dangerous assorted wastewater and compound contaminations counting aliphatic, fragrant hydrocarbons, heterocyclic and other nitrogen and sulphur segments, are dealt with to decrease their gathering in consumption water.

This is a framework that procedures wastewater from a modern unit, for example, a processing plant. Similarly as with the home framework, the limits reach out from the wastewater source (the modern procedures) through to the advances and forms for restoring the prepared wastewater to the biological community. Figure 3 shows that disentangled representation of this. It contrasts from the home framework as far as:

- The sorts of advancements creating the waste
- The way wastewater is overseen at the source
- The sort of waste delivered
- Chemical and metals.

The framework is comparable in that it incorporates individuals, reusing, the treatment innovations and the environment inside which the business sits.

2.1.3 Business Wastewater

Business wastewater is the wastewater produced dominatingly from commercial or business focuses. It establishes manure from sterile offices, as well as strong squanders and wastewater (consolidated) beginning from business focuses. More than 60 for each penny of the heterogeneous sewage are create from eateries; clothing business focuses and benefits stations, nightclubs, off licenses and so forth, which do not experience pre-treatment afore transfer.

2.1.4 Tempest Water

The main part of municipal water run-off created from various sources is water and released into drain frameworks or accepting water bulks. It establishes a blend of water, dregs, or largely strong discarded constituents, run of the mill of urban water run-offs or streams. The large precipitation miserable bumpy slants clears laterally strong discarded constituents and free soil constituents, through little dump downstream. Situations happen were these materials develop and stop up the streambeds causing flooding. The diverse sort of parts show in wastewater was shown in Table 2.

3 Sustainability Wastewater Management

Customarily, water asset administration has emphatically depended on wastewater treatment to guarantee water quality is kept up. In any case, the noteworthy political ways to deal with overseeing water quality are under new weights, for example, micro pollution. Various micro pollutants are not defenceless against current treatment and are in this manner relentlessly transported into the amphibian condition. There remains a lot of vulnerability concerning the capacity of innovative treatment advances, for example, ozonation or enacted carbon, to channel micro pollutants and their expanded vitality needs and expenses.

To exhibit which choices there are to taking care of water quality issues, we take a gander at approach instruments beforehand acquainted with handle "customary" water quality issues, (for example, full scale contamination) and give a review of potential future strategy plan and arrangements [12]. Keeping in mind the end goal to comprehend which of these arrangement instruments are most suited to directing new wonders, for example, micro pollution, break down the attributes of micro pollution as an approach issue, i.e., its circumstances and results, and distinctive manageability measurements (e.g., long haul, multi-level). Nonetheless, this investi-

Table 2 Diverse sort of parts show in wastewater

S. no.	Constituents	Special interest	Environmental Effects
1	Micro-organisms	Pathogenic bacteria, virus and worm eggs	Poses risk when bathing and eating of shellfish
2	Biodegradable organic materials	Oxygen depletion in rivers, and lakes	Causes changes in aquatic life
3	Other organic materials	Detergents, pesticides, fat, oil and grease, colouring, solvents etc.	Causes toxic effect, aesthetic inconveniences, and bio-accumulate
4	Nutrients	Nitrogen, phosphorus, ammonia	Causes eutrophication and toxic effect
5	Heavy metals	Hg, Pb, Cd, Cr, Cu, Ni	Causes corrosion and toxic effect
6	Other inorganic materials	Acids, e.g. H_2S, bases	Causes corrosion, and toxic effect
7	Thermal effects	Hot water	Change living conditions of flora and fauna
8	Odour (taste)	Hydrogen sulphide	Aesthetic inconveniences, toxic effect
9	Radioactivity	Pose toxic effect, and accumulate	–

gation has demonstrated that there is a need to outline a reciprocal instrument blend that incorporates specialized arrangements and source-guided strategy instruments to diminish the utilization of micro pollutants before they enter waters. Notwithstanding, the presentation of source-coordinated measures in the past has demonstrated that they run as one with behavioural changes of target-gatherings, which makes their presentation a testing undertaking. Target gatherings may wish to minimize expenses and adaptability in activities high [13].

Breaking the issue into littler parts would be one method for permitting focused on reactions and expanding agreeableness by the objective gathering. To do as such, flat coordination among various approach fields is required. Moreover, and like the larger part of water contamination issues, micro pollution does not stop at national fringes. Thusly, productive and powerful instrument decisions can just accomplish their objectives in the event that they are outline in a universal and trans-limit setting.

This segment quickly audits the idea of incorporated wastewater frameworks administration. While picking how to oversee wastewater and squanders in your group, a qualification made between the 'framework' and the specialized building arrangements that may be utilize inside that framework. A wastewater framework will incorporate advancements; however will likewise incorporate the procedures that happen inside and between the diverse innovative parts [14]. Wastewater frameworks likewise incorporate individuals and their activities and conduct, and additionally the regular biological community forms inside which the advancements work.

Overseeing wastewater at source

Choices for administration at the source include:

• Water-sparing practices in and around the home
• Decision of family unit items that will enter the wastewater stream.

The measure of water utilized by a group will be a noteworthy factor in choosing the measure of a wastewater framework. Reasonably clearly, water protection can decrease the measure of wastewater that can manage. It is conceivable to figure whether water protection will influence the framework plan and last expenses. The outline of a wastewater framework should likewise assess what materials are going down the channels. The nearness of distinctive poisonous materials may request a more elevated amount of treatment than would regularly happen. What goes down the deplete additionally has a tremendous effect on how well septic tanks and on location frameworks work. Once more, a full-framework audit, which deals with the sum of harmful materials, oils, fats, oils, and so on going down the deplete, will impact the outline of the last framework.

4   Characteristics of Wastewater

Anyone of water is equipped for acclimatizing a specific measure of poisons without genuine impacts due to the weakening and self-purging variables, which are available. On the off chance extra contamination happens, the nature of the getting water will be adjust and its appropriateness for different utilizations be hindered. Comprehension of the impacts of contamination and the control measures that are accessible is along these lines of impressive significance to the productive administration of water assets. Metropolitan wastewater comprises of a blend of broke down, colloidal, and particulate natural and inorganic materials. Metropolitan wastewater contains 99.9% water. The remaining materials incorporate suspended and broke down natural and inorganic issue as well as microorganisms. These materials make up the physical, concoction, and natural qualities that are attributes of private and mechanical waters [15]. The physical nature of wastewater is detailed as far as its temperature, shading, and turbidity. The temperature of wastewater is somewhat higher than that of the water supply. This is an essential parameter due to its impact upon oceanic life and the dissolvability of gasses. The temperature shifts marginally with the seasons, ordinarily staying higher than air temperature amid the majority of the year and coming up short lower as it were amid the sweltering summer months.

The shade of a wastewater is generally characteristic of age. New water is normally dim, septic wastewater confers a dark appearance. Scents in wastewater are caused by the decay of natural issue that produces hostile noticing gasses, for example, hydrogen sulphide. Wastewater scent largely can give a relative sign of its condition. Turbidity in wastewater is caused by a wide assortment of suspended solids. Suspended solids are characterized as the material that can expelled from water by filtration through arranged films. Compound qualities of wastewater are communicated as far as natural and inorganic constituents. Diverse synthetic investigations outfit valuable and particular data with regard to the quality and quality of wastewater. Natural mixes in the wastewater are the hugest factor in the contamination of numerous normal waters. The central gatherings of natural issue found in metropolitan wastewater are proteins, sugars, and fats and oils. Sugars and proteins are effectively biodegradable [16].

Fats and oils are steadier and can deteriorate by microorganisms. In expansion, wastewater may likewise contain little division of engineered cleansers, phenolic mixes, and pesticides and herbicides. These mixes, contingent upon their fixation, may make issues, for example, non-biodegradability, frothing, or cancer-causing nature. The groupings of these poisonous natural mixes in wastewater are small. Their sources are generally mechanical squanders and surface overflow. The inorganic mixes most found in wastewater are chloride, hydrogen particles, alkalinity-causing mixes, nitrogen, phosphorous, and sulphur mixes, and overwhelming metals. Follow groupings of these mixes can fundamentally influence creatures in the getting water through their developing restricting attributes. The quality and types of smaller scale and plainly visible plants and creatures, which make up the natural attributes in an accepting water body, considered as the last trial of wastewater treatment viability. Inside the treatment office, the wastewater gives the ideal medium to microbial development, regardless of whether it is oxygen consuming or anaerobic. Microorganisms and protozoa are the keys to the natural treatment process utilized at most treatment offices, and to the normal natural cycle in getting waters. In the nearness of adequate disintegrated oxygen, microorganisms change over the solvent natural issue into new cell tissues, carbon dioxide and water.

5   Wastewater Treatment Technologies

The critical increment of municipal wastewater and arrival of unprocessed wastewater increments in creating countries implies the essential to give essential, auxiliary and additionally propelled treatment to energize re-utilized is very basic. Studies show that numerous tropical nations are presently putting resources into modest and supportable little scale and minimal effort (instead of customary) sewage treatment advancements for sewage treatment.

5.1   Treatment Methodologies

The partial reduction or complete removal of excessive impurities contain in wastewater. Usually wastewater treatment is mean the reduction or removal of the solids from wastewater since impurities are generally because of the nearness of solids in wastewater.

5.1.1 Physical Unit Operations

Physical unit operation and these are gone for evacuating the inorganic solids generally and natural solids to some degree from wastewater. Indeed, it involves what is known as a strong fluid partition process that is the sedimentation. Since it is a gravitational procedure the physical laws represents the procedure so that can utilize the physical laws. That is traditional mechanics can be utilized for outline of framework to evacuate the solids from the fluids like sedimentation tank etc.

Amid the essential wastewater treatment process, wastewater is incidentally held in a tank where heavier solids can settle to the base, while any lighter solids and rubbish buoy to the surface. The settled and skimming materials are desludged or kept down and the staying fluid released or put through an optional treatment process. Expert gear is have to evacuate essential ooze that has settled as natural and inorganic solids on the base of settling tanks. Skimming material likewise needs to expel. Essential wastewater treatment plants are configuration to use in conjunction with our natural treatment modules, where settlement or capacity is required. Every essential wastewater tank is accessible with a decision of wellbeing highlights and desludging hardware that can chose by the necessities of an individual site. A portion of the cases of physical unit frameworks are screening, coarseness expulsion and sedimentation tank.

(a) **Screening**

Screening is the primary line of treatment at the passageway to the wastewater treatment plant where six new fine screens, organized in parallel channels, block strong material in the influent wastewater. The screening component may comprise of parallel bars, bars or wires, wirework, or punctured plate, and the openings might be of any shape yet for the most part are round or rectangular spaces. A screen made out of parallel bars or poles is frequently called a "bar rack" or a coarse screen and is utilized for evacuation of coarse solids [17]. A screen is a gadget with openings, largely of uniform size, that is use to hold solids found in the influent wastewater to the treatment plant. The schematic graph of screening process was shown in Fig. 4.

The guideline part of screening is to expel coarse materials from the stream that could:

(i)   Damage resulting process hardware
(ii) Reduce general treatment process unwavering quality and viability
(iii) Contaminate conduits.

Fine screens are at times utilized as a part of place of or following coarse screens where more noteworthy expulsions of solids are required

(i)   Protect process gear or

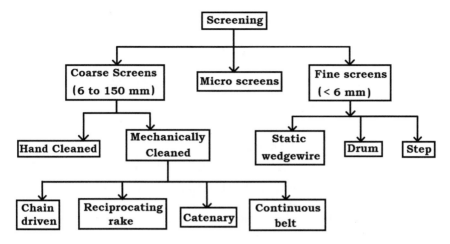

Fig. 4 Schematic diagram of screening process

(ii) Eliminate materials that may repress the valuable reuse of bio solids.

(b) **Grit chamber**

Sand, slag, Cinder, Bone Chip, eggshells, and so forth, of size under 0.2 mm is incorporated into coarseness. It is not putrescible and has higher subsidence esteem than the natural issue. It is consequently conceivable to expel coarseness from the wastewater effectively by lessening the wastewater speed in long channel called as coarseness channel. The speed is diminish to around 0.3 m/s. The settled coarseness is wash before its transfer.

Coarseness Chamber is accommodated the reason for evacuation of residue and sand particles mostly so a similar will not cause the wear and tear of vanes of pumps, stopping up of funnels, and additionally, and valve operation troublesome. Evacuation of coarseness additionally lessens aggregation of dormant material in consequent treatment units. Establishing impacts are counteracted in settling tanks and digester by expulsion of coarseness. Coarseness Removal Unit might be a coarseness Channel, Grit Chamber or a Grit Basin.

Coarseness chambers are configuration to expulsion coarseness, comprising of sand, rock, ashes, or other overwhelming strong materials that have dying down speeds or particular gravities considerably more prominent than those of the natural solids in wastewater [18]. Coarseness chambers are most normally situated after the bar screens and before the essential sedimentation tanks. Essential sedimentation tanks work for the evacuation of overwhelming natural solids. In a few establishments, coarseness chambers go before the screening offices. Largely, the establishment of screening offices in front of the coarseness chambers makes the operation and support of the coarseness evacuation offices less demanding.

Grit chambers are provided to:

- Protect moving mechanical hardware from scraped spot and going with strange wear;
- Reduce arrangement of substantial stores in pipelines, and channels, and courses; and
- Reduce the recurrence of digester cleaning caused by unreasonable gatherings of coarseness.

There are two kinds of coarseness chambers.

(i) Horizontal flow Grit Chambers

In level stream compose, the stream goes through the chamber in a flat heading and the measurements of the unit, an influent dispersion entryway, and a weir at the emanating end control the straight-line speed of stream. Rectangular or square flat stream coarseness expulsion have utilized for a long time. Rectangular flat stream coarseness chambers are the most seasoned sort of coarseness chamber utilized. It is speed controlled write. These units intended to keep up a speed as near 0.3 m/s as pragmatic and to give adequate time to coarseness particles to settle to the base of the channel. The outline speed will help most natural particles through the chamber and will have a tendency to suspend any natural particles that settle however will allow the heavier coarseness to settle out.

(ii) Aerated Grit Chamber

The circulated air through sort comprises of a winding stream air circulation tank, where the winding speed is incited and controlled by the tank measurements and amount of air provided to the unit. In circulated air through coarseness chambers, air is presented along one side of a rectangular tank to make a winding stream design opposite to the move through the tank. The heavier coarseness particles that have higher settling speeds settle to the base of the tank. Lighter, basically natural, particles stay in suspension and go through the tank. The speed of roll or fomentation represents the span of particles of a given particular gravity that will evacuate. On the off chance that the speed is excessively awesome, coarseness will be does of the chamber; in the event that it is too little, natural material will be expel with the coarseness. Luckily, the amount of air is effortlessly balanced, just about 100% evacuation will acquire and the coarseness will be well wash. Coarseness that is not well wash and contains natural issue is a scent aggravation and pulls in creepy crawlies.

(c) Flow equalization

Stream balance is a strategy used to conquer the operational issues caused by stream rate varieties, to enhance the execution of the downstream procedures, and to decrease the size and cost of downstream treatment offices. Stream balance essentially is the damping of stream rate varieties to accomplish a steady or about consistent stream rate and can connected in various distinctive circumstances, contingent upon the attributes of the accumulation framework. The primary applications are for the balance of

- Dry-climate streams to lessen crest streams and loads,

- Wet-climate streams in clean accumulation frameworks encountering inflow and invasion, or
- Combined storm water and clean framework streams.

Application of flow equalization:

- Biological treatment is improved, in light of the fact that stun loadings are dispense with or can limited, hindering substances can be weakened, and pH can be balanced out.
- The gushing quality and thickening execution of optional sedimentation tanks following natural treatment is enhanced through enhanced consistency in solids stacking.
- Effluent filtration surface are prerequisites are diminish, channel execution is enhanced, and uniform channel backwash cycles are conceivable by bring down pressure driven stacking.
- In compound treatment, damping of mass stacking enhances synthetic bolster control and process unwavering quality.

Impediments of stream evening out:

- Relatively expansive land regions or destinations are require.
- Equalization offices may must be cover for smell control close neighborhoods.
- Additional operation and upkeep is required.
- Capital cost is expanded.

(d) **Primary Sedimentation**

Two parallel, chain-driven flight scrubbers gather the muck. These move ceaselessly along the inclining floors of the tanks, gradually furrowing the ooze towards the finish of the tank where a cross gatherer (likewise chain and flight) moves the slop into a profound container. From here, new divergent pumps to an ooze sump evacuate it. Fan-molded water planes guide filth, which ascends to the surface of the tanks, to the delta end of the tank. Here, it is lifted over a divider and into a trough by turning rubbish authorities and conveyed into the muck sump. The slop and filth from the essential sedimentation tanks is pump to the gravity thickeners. After the ooze has thickened in the gravity thickeners, it sent to the gravity belt thickeners for additionally thickening before sent to the digesters. The goal of treatment by sedimentation is to evacuate promptly settle capable solids and gliding material and in this manner lessen the suspended solids content. Essential sedimentation can use as a preparatory advance in the further handling of the wastewater. Proficiently composed and worked essential sedimentation tanks should expel from 50 to 70% of the suspended solids and from 25 to 40% of the BOD. All treatment plants utilize mechanically cleaned sedimentation tanks of institutionalized roundabout or rectangular outline. At least two tanks ought to be give with the goal that the procedure may stay in operation while one tank is out of administration for support and repair work [19].

Sedimentation tanks are normally design based on a surface loading rate expressed as cubic meters per square meter of surface area per day,

m^3/m^2d. The determination of a reasonable stacking rate relies upon the sort of suspension to partition. At the point when the region of the tank has built up, the detainment period in the tank is oversee by water profundity.

(e) Odour Removal

Odour control is a vital part of the wastewater treatment process. Smelly air is gathered at different phases of treatment by ventilation fans and ducted to promoter fans, which go it through earth channels (bio channels). Smelly air can uniformly disseminated, underneath the media by an arrangement of header and dispersion funnels. As it permeates upwards, microscopic organisms inside the media treat the musty mixes [20]. Physical and bacterial procedures expel putrid mixes before released to air. Bio channels likewise treat air separated from different zones of the treatment plant including the pre-treatment blending chamber, gravity thickeners, the splitter boxes and the bio solids dewatering building.

5.1.2 Chemical Unit Processes

Chemicals can be utilized amid wastewater treatment in a variety of procedures to speed up purification. These synthetic procedures, which actuate compound responses, substance unit forms, and utilized nearby organic and physical cleaning procedures to accomplish different water benchmarks. There are a few unmistakable compound unit forms, including synthetic coagulation, concoction precipitation, substance oxidation and propelled oxidation, particle trade, and compound balance and adjustment, which can connected to wastewater amid cleaning.

(i) Chemical precipitation

Chemical precipitation is the most known technique for removing separated metals from wastewater game plan containing unsafe metals. To change over the deteriorated metals into solid particle shape, a precipitation reagent can added to the mix. An engineered reaction, enacted by the reagent, makes the broke down metals outline solid particles. Filtration would then have the capacity to use to oust the particles from the mix. How well the methodology capacities is dependent upon the kind of metal present, the centralization of the metal, and the kind of reagent used. In hydroxide precipitation, a regularly used compound precipitation process, calcium or sodium hydroxide can used as the reagent to make solid metal hydroxides. In any case, it can be difficult to make hydroxides from separated metal particles in wastewater in light of the way that various wastewater courses of action contain mixed metals.

(ii) Chemical coagulation

This creation methodology incorporates destabilizing wastewater particles so they add up to in the midst of substance flocculation. Fine solid particles scattered in wastewater pass on negative electric surface charges (in their conventional stable state), which shield them from forming greater get-togethers and settling. Substance coagulation destabilizes these particles by showing insistently charged coagulants that by then reduction the negative particles' charge. Once the charge can diminished, the particles straightforwardly outline greater social events. Next, an anionic flocculant is familiar with the mix. Either since the flocculant responds against the emphatically charged blend, kills the molecule gatherings or makes connects between them to tie the particles into bigger gatherings [21]. After bigger molecule bunches are framed, sedimentation can be utilized to expel the particles from the blend.

(iii) **Compound oxidation and propelled oxidation**

With the introduction of an oxidizing administrator in the midst of substance oxidation, electrons move from the oxidant to the defilements in wastewater. The defilements by then experience helper change, ending up less ruinous blends. Essential chlorination uses chlorine as an oxidant against cyanide. By the by, dissolvable chlorination as an engineered oxidation process can provoke the making of harmful chlorinated blends, and additional advances may be required. Impelled oxidation can help remove any regular disturbs that are conveyed as a reaction of substance oxidation, through methods, for instance, steam stripping, air stripping, or incited carbon adsorption.

(iv) **Ion exchange**

Exactly when water is too hard, it is difficult to use to clean and consistently leaves a diminish store. (This is the reason pieces of clothing washed in hard water much of the time holds a filthy tint.) A molecule exchange process, similar to the switch osmosis process, can use to assuage the water. Calcium and magnesium are standard particles that incite water hardness. To unwind the water, firmly charged sodium particles can displayed as deteriorated sodium chloride salt, or salty water [22]. Hard calcium and magnesium particles exchange places with sodium particles, and free sodium particles can released in the water. By the by, ensuing to softening a great deal of water, the softening course of action may stack with excess calcium and magnesium particles, requiring the plan restored with sodium particles.

(v) **Chemical stabilization**

This strategy works in a similar way as mixture oxidation. Overflow can treated with a ton of a given oxidant, for instance, chlorine. The introduction of the oxidant backs off the rate of common advancement inside the sludge, and circulates air through the mix. The water can oust from the overflow. Hydrogen peroxide can in like manner be used as an oxidant, and may be a more down to earth choice.

5.1.3 Biological Unit Processes

Biological treatment forms are those that utilization microorganisms to coagulate and expel none settle capable colloidal solids to settle the natural issue. Organic squander water medications can used to expel disintegrated and colloidal natural issue in a waste. The emanating from the essential sedimentation tank contains about

60–80% of the flimsy natural issue initially display in sewage. This colloidal natural issue, which passes the essential clarifiers, without settling there, must expelled by facilitate treatment. This further treatment of sewage is called auxiliary treatment in which organic and compound forms can utilized to evacuate the majority of the natural issue [23]. The optional treatment is coordinated mainly towards the expulsion of decomposable organics and suspended solids. It involves 99.9% water and 0.1% solids.

Biological treatment forms are those that utilization microorganisms to coagulate and expel the non settle able colloidal solids to settle the natural issue. Unique high-impact natural medicines are as per the following:

(i) **Activated sludge process**

The term activated sludge is utilized to show the ooze which is gotten by settling sewage in nearness of copious oxygen. The activated sludge is naturally dynamic and it contains an incredible number of oxygen consuming microbes and different microorganisms which have a got a bizarre property to oxidize the natural issue. The initiated ooze process has utilized widely all through the world in its traditional shape and changed shapes, which are all fit for meeting optional treatment gushing points of confinement. It incorporates preparatory treatment comprising of bar screen as a base and, as required, comminutor, coarseness chamber, and oil and oil evacuation units [24].

In enacted slime process wastewater, containing regular issue is flowed air through in an air course bowl in which littler scale living creatures use the suspended and dissolvable characteristic issue. Some bit of common issue can mix into new cells and part can oxidize to CO_2 and water to decide essentialness. In started slop systems, the new cells formed in the reaction can removed from the liquid stream in the state of a wooly slop in settling tanks. A bit of this settled biomass, delineated as incited slop is returned to the air flow tank and whatever is left of the structures waste or plenitude filth.

(ii) **Trickling filter**

The sewage is permitted to sprinkle or to stream over a bed of coarse, harsh, hard material and it is at that point gathered through the under drainage framework. The oxidation of the natural issue can complete under oxygen consuming conditions. A bacterial film known as a bio-film can framed around the particles of separating media and for the presence of this film, the oxygen is provided by the discontinuous working of the channel and by the arrangement of appropriate ventilation offices in the body of the channel [25]. The shade of this film is blackish, greenish and yellowish. It comprises of microscopic organisms, parasites, green growth, lichens, protozoa, and so forth.

Types of Trickling Filters

The trickling filters are broadly divided into two categories:

• Standard rate trickling filters

- High-rate or high capacity trickling filters.

(iii) Rotating biological contactors

Rotating Biological Contactors can utilized to treat in a financially perceptive way from 5000 gallons to a large number of gallons every day of household and modern wastewaters. The RBC process gives an amazingly high level of treatment giving emanating qualities as low as 5 mg/L of dissolvable Biochemical Oxygen Demand (BOD) and 1 mg/L alkali nitrogen. They can utilized for essentially bringing down the levels of dissolvable organics and Chemical Oxygen Demand (COD).

(iv) Aerobic treatment

Oxygen consuming treatment has used to clear take after normal precarious blends in water. It has furthermore been used to trade a substance, for instance, oxygen, from air or a gas organize into water in a methodology called "gas adsorption" or "oxidation", i.e., to oxidize press and additionally manganese. Air dissemination similarly gives the escape of separated gasses, for instance, CO_2 and H_2S. Air stripping has furthermore utilized suitably to oust NH_3 from wastewater and to clear eccentric tastes and other such substances in water [26]. Energetic treatment with bio misuses is effective in diminishing ruinous vaporous outpourings as nursery gasses (CH_4 and N_2O) and noticing salts.

(a) Oxidation ponds

Oxidation lakes are incredible systems where the oxygen required by the heterotrophic microorganisms (a heterotroph is a living being that cannot settle carbon and utilizations natural carbon for development) can give by exchange from the environment as well as by photosynthetic green growth. The green growth are confined to the euphotic zone (daylight zone), which is frequently just a couple of centimetres profound. Lakes are built to a profundity of near 1.2 and 1.8 m to guarantee most extreme infiltration of daylight, and seem dull green in shading because of thick algal advancement. In oxidation ponds, the green growth utilize the inorganic mixes (N, P, CO_2) discharged by high-impact microscopic organisms for development utilizing daylight for vitality. They discharge oxygen into the arrangement that thusly can used by the microscopic organisms, finishing the harmonious cycle. There are two specific zones in facultative lakes: the upper oxygen-expending zone where bacterial (facultative) activity happens and a lower anaerobic zone where solids settle out of suspension to outline an ooze that can debased anaerobically.

(b) Aerated Lagoons

Circulated air through tidal ponds are huge (3–4 m) stood out from oxidation lakes, where oxygen is given by aerators yet not by the photosynthetic development of green development as in the oxidation lakes. The aerators keep the microbial biomass suspended and give satisfactory separated oxygen that grants maximal incredible development. On the other hand, bubble air flow can for the most part use where the air pockets can created by stuffed air pumped through plastic tubing laid through the base of the lagoon. A predominately-bacterial biomass makes and, while there is

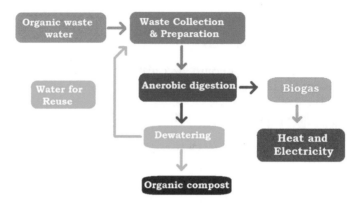

Fig. 5 Anaerobic wastewater treatment

neither sedimentation nor ooze re-establish, this strategy depends on agreeable mixed liquor surrounded in the tank/lagoon. In this manner, the airflow lagoons are sensible for strong yet degradable wastewater, for instance, wastewaters of support wanders. The pressure driven maintenance time (HRT) ranges from 3 to 8 days in view of treatment level, quality, and temperature of the influent. Largely, HRT of around 5 days at 20 °C accomplishes 85% evacuation of BOD in family unit wastewater. In any case, if the temperature falls by 10 °C, at that point the BOD expulsion will lessen to 65%.

(v) **Anaerobic treatment**

The anaerobic treatment can realized to treat wastewaters rich in biodegradable normal issue and for energize treatment of sedimentation sludges. Solid natural wastewaters containing many biodegradable materials are released for the most part by horticultural and sustenance preparing ventures. These wastewaters are hard to deal with vigorously because of the inconveniences and costs of satisfaction of the raised oxygen request to protect the high-impact conditions. Interestingly, anaerobic debasement happens without oxygen. In spite of the fact that the anaerobic treatment is tedious, it has a huge number of points of interest in treating solid natural wastewaters. These preferences incorporate hoisted levels of filtration, inclination to deal with high natural burdens, creating little measures of slimes that are typically exceptionally steady, and generation of methane (latent ignitable gas) as result.

Anaerobic digestion is a complex multistep process with respect to science and microbiology. Normal materials are degraded into crucial constituents, finally to methane gas under the nonappearance of an electron acceptor, for instance, oxygen [27]. Fitting wastewaters fuse trained creatures fertilizer, sustenance taking care of effluents, oil misuses (if the toxic quality is controlled), and canning and dyestuff misuses where dissolvable common issues are executed in the treatment. Schematic outline of anaerobic wastewater treatment was shown in Fig. 5.

Most anaerobic procedures (solids maturation) happen in two foreordained temperature ranges: mesophilic or thermophilic. The temperature ranges are 30–38 °C and 38–50 °C, individually. As opposed to oxygen consuming frameworks, total adjustment of natural issue is not achievable under anaerobic conditions. Hence, consequent vigorous treatment of the anaerobic effluents is generally basic. The last waste issue released by the anaerobic treatment incorporates solubilised natural issue that is submissive to high-impact treatment exhibiting the likelihood of introducing aggregate anaerobic and oxygen consuming units in arrangement.

6 Sustainability

The term sustainability is critical to actualize wastewater frameworks. Manageability identifies with five angles as characterized by the Sustainable Sanitation Alliance. Sanitation in this regard incorporates wastewater administration and release too. The fundamental target of a sanitation and wastewater treatment framework is to secure and advance human wellbeing by giving a spotless domain and breaking the cycle of sickness. With a specific end goal to be feasible a sanitation framework must be not just financially reasonable, socially adequate, and in fact and institutionally proper, it ought to ensure the earth and the normal assets. While progressing a current or potentially outlining another sanitation framework, maintainability criteria identified with the accompanying perspectives ought to considered:

(i) **Health and cleanliness**:

Incorporates the danger of presentation to pathogens and unsafe substances that could influence general wellbeing at all purposes of the sanitation framework from the can by means of the accumulation and treatment framework to the point of re-utilize or transfer and downstream populaces.

(ii) **Environment and normal assets**:

Includes the required vitality, water and other regular assets for development, operation and support of the framework, and additionally the potential outflows to the condition coming about because of utilization. It additionally incorporates the level of reusing and re-utilize honed and the impacts of these (e.g. recycling wastewater; returning supplements and natural material to horticulture), and the ensuring of other non-inexhaustible assets, for instance through the generation of sustainable vitalities (e.g. biogas).

(iii) **Technology and operation**:

Joins the usefulness and the simplicity with which the whole framework counting the accumulation, transport, treatment and re-utilize as well as definite transfer can be developed, worked and, checked by the nearby group or potentially the specialized groups of the neighbourhood utilities more, the vigour of the framework, its power-lessness towards control cuts, water deficiencies, surges, and so on. In addition, the adaptability and versatility of its specialized components to the current framework and to statistic and financial advancements are vital angles to assess.

(iv) **Financial and monetary issues**:

Identify with the limit of family units and groups to pay for sanitation, counting the development, operation, support and important reinvestments in the framework.

(v) **Socio-social and institutional viewpoints**:

The criteria in this classification assess the socio-social acknowledgment furthermore, fittingness of the framework, accommodation, framework recognitions, sexual orientation issues and effects on human poise, consistence with the legitimate structure and steady and effective institutional settings.

6.1 Sustainability Analysis of Wastewater Treatment Systems

In prior circumstances and even today, architects and lawmakers about dependably utilize a basic cost/advantage investigation while picking a wastewater framework. This implies, for occurrence, just the release of natural issue (BOD) or phosphorus and the cost can looked upon. In any case, the journey for manageability is vital in light of the fact, numerous issues are coming like an unnatural weather change, fermentation, lessening ozone layer, smaller scale natural poisons and other poisonous concoction matters, eutrophication, reducing vital assets like phosphorus, potassium and oil and different dangers to humanity, verdure. This demonstrates numerous markers must utilized when choosing what sort of wastewater frameworks need to executed. In addition, ought to pick the wastewater framework that contributes most to a general maintainable future.

The thought supportability ought to incorporate nature; economy and sociological viewpoints and the manageability should likewise perform on three distinct stages:

(i)   Nearby, where sterile and wellbeing perspectives are of worry in time sizes of hours or days.
(ii) Local, where exemplary ecological issues work in time sizes of months or, on the other hand years.
(iii) Worldwide, where maintainability matters in a period size of decades or hundreds of years.

6.2 Economic Sustainability

Economic maintainability suggests an arrangement of generation that fulfils show utilization levels without trading off future needs. The 'sustainability' that 'economic sustainability' looks for is the 'supportability' of the financial framework itself. Generally, financial specialists, accepting that the supply of normal assets was boundless, put undue accentuation on the limit of the market to dispense assets effectively. They likewise accepted that monetary development would bring the mechanical ability to recharge characteristic assets obliterated in the creation procedure. Today, nevertheless, an acknowledgment has developed that characteristic assets are not interminable. The developing size of the monetary framework has stressed the regular asset base.

A financial framework outlined in light of the hypothesis of 'economic sustainability' is one compelled by the necessities of 'ecological sustainability'. It limits asset use to guarantee the 'sustainability' of regular capital. It does not try to accomplish 'economic sustainability' at the cost of 'ecological sustainability'.

6.2.1 Operation and Administration

Working and upkeep costs related with wastewater treatment incorporate work, vitality, and buy of chemicals and substitution gear. This can be credited to additional much automated gear and complex procedures that require impressive vitality inputs.

6.2.2 Client Costs

Wastewater treatment costs are largely dependent upon the sort of treatment development, its viability, and the discharge decision used. Another factor is the people assess served.

6.3 Ecological Sustainability

6.3.1 Energy Use

A larger part of process and support expenses might credited to vitality utilization amid air circulation and pumping of water and solids. Some enacted ooze frameworks may have bring down vitality utilization as a result of inside vitality ignition of methane gas delivered in-house, especially from anaerobic absorption. In future investigations, vitality creation related to squander administration could incorporated as a property of a maintainable innovation. Vitality utilize is frequently connected with worldwide ecological issues, for example, carbon dioxide emanations. For instance, an actuated muck framework serving a populace of 1000 individuals can possibly deliver up to 1400 ton of CO_2 for process furthermore, 50 ton of carbon dioxide for upkeep over a 15-year life.

Different open doors exist in lessening vitality utilize and related effects. These incorporate the kind of wastewater treatment innovation chose; utilization of reused materials for development; rectify estimating and rating of hardware for operation, particularly to pump necessities; and reuse of waste total from annihilation. Plant configuration can likewise all the more painstakingly join issues of vitality preservation also, as specified already, utilization of in-house methane generation may diminish outer vitality needs.

6.3.2 Expulsion of Water Quality Constituents

The normal water quality constituents related with wastewater treatment are BOD, TSS, phosphorus, nitrogen and fecal coliforms. There are some real complexities in clearing efficiencies of each treatment advancement. These exuding qualities in the end choose whether advance treatment is required, what sort of discharge choices can be used and specifically their potential for reuse. Fate and removal of unsafe and deadly blends is a generally recognized test in wastewater treatment and markers that pass on information on transmissions or floods of these blends are as often as possible proposed. If wastewater treatment is to be sensible in the whole deal, better waste organization techniques may need to make that join close-by money related development, take out the usage and despicable exchange of family unsafe waste, and unite as one with prosperity providers that suggest pharmaceutical chemicals.

The utilization of boundaries at different focuses in wastewater treatment frameworks can have a colossal effect in overseeing risky substances. Besides, expulsion efficiencies of pathogens, overwhelming metals, and other lethal mixes have significant ramifications on water reuse plans. The particular sort of wastewater reuse activities eventually characterizes the nature of wastewater required and the ensuing treatment forms expected to accomplish this quality.

6.4   Societal Sustainability

6.4.1   Open Support

Open cooperation is regularly a disregarded angle when choosing the most proper wastewater treatment innovation for a specific group. While a few controls assign a particular innovation through a "best innovation" process, the recognitions and inclinations of the open for the choice and usage of a specific innovation is critical if innovation is to be coordinated with neighbourhood and more extensive manageability concerns.

The components considered essential in choosing a manageable treatment framework will fluctuate from group and area due to land and demographical substances particular to a territory. Regardless, the components right now utilized as a part of wastewater treatment choice are more often than not execution and moderateness. In created nations, a wastewater treatment plant's proficiency, dependability, slop transfer and land prerequisites can viewed as basic over reasonableness. Subsequently there is an inclination to select mechanical frameworks over option treatment frameworks. In creating nations, reasonableness and the suitability of the innovation can viewed as basic. In this manner, these nations regularly select basic, financially perceptive proper innovation, over more automated innovation. Choosing an advanced treatment framework for a group with low-salary families may put undue money related hardship on them.

6.4.2   Community Estimate Served

The span of a group can direct the sort of treatment framework chose, its ability, and thus its maintainability. Increment in populace frequently implies a bigger plant limit is required. Mechanical and tidal pond frameworks are fit for overhauling a bigger populace than arrive treatment frameworks. Nevertheless, mechanical frameworks can regularly pick over tidal pond frameworks to benefit these populaces. An integral factor in picking mechanical frameworks that serve expansive populaces, particularly in urban regions is the land prerequisite or open space accessibility.

Metropolitan contamination loadings can related with urban territories because of their expansive material sources of info and yields contrasted with littler groups. This may make a weight on the encompassing condition to which disintegrated and strong residuals are returned in light of the fact that the encompassing has restricts on how much poison stacking it can acknowledge. Appropriately, if wastewater frameworks are to be maintainable, at that point contemplations of material adjusts, especially water and compound transitions, are required to keep up a legitimate adjust of supplements in nature; staying away from the gathering of poisons in a single environment or, on the other hand lack of supplements in another.

6.4.3   Annoyance from Smell

Wastewater treatment workplaces, paying little personality to how much arrange eventually may make fragrance as reactions of the wastewater social occasion and treatment process. The closeness of smell in any wastewater treatment office is commonly an in vogue issue that regularly motivates open mediation and eventually managerial association affiliation. All the treatment structures can possibly convey spoiled releases. Land treatment structures have the slightest fragrance potential than mechanical and lagoon systems, due to pre-treatment of the wastewater before arrive application. Fragrance issues may in like manner rise, if considerable solids and green development have not been emptied going before territory treatment. Govern notice issues ordinarily occur at pumping stations, delta and outlet directing, and sewer vents if any are accessible. Scents from lagoon structures may in like manner be a direct result of over-troubling or extreme surface junk that has allowed to total.

6.5   Significance of Sustainability

The term 'sustainability' alludes to abroad idea, including different interrelated parameters in regards to the earth, individuals and vitality assets. The hugeness of sustainability for manufactured situations can well know as a multi-disciplinary approach to the pondering of natural, monetary and socio-social concerns. It is to moderate the negative natural effects and to fit the living situations with financial examples.

7   Environment Re-entry or Re-utilize

7.1   Re-entry of Treated Waste into the Biological System

Treated wastewater might be come back to the biological community through direct guide release toward a water body, for example, a stream, lake, wetland or estuary, or to ocean. Then again, the treated wastewater might be come back to arrive by different water system strategies, for example, surge water system, overhead sprinklers or sub-surface drippers. Choices for restoring the treated wastewater to the biological community inside the site limits (regularly alluded to as on location transfer) depend particularly on the site's qualities, for example, soil sorts, zone and incline of land accessible, area of groundwater, and neighbourhood atmosphere. Choices incorporate drainage into the dirt sub-surface, water system (surface or sub-surface) and transpiration.

7.1.1   Gasses

These incorporate gasses, for example, smelling salts, methane and hydrogen sulphide, and musty natural gasses, for example, mercaptans, indole and skatole. These can re-enter at different focuses, for example, if water turns septic from an overburden of natural material, or at the point slop can landfilled. Methane can develop inside a site and should figure out how to decrease dangers to encompassing properties. Hazard administration and site administration gets ready for landfills to oversee flammable gasses and smell will be an essential some portion of the re-entry procedure. Frequently people group do not factor in the expenses of landfill administration into wastewater administration costs while picking choices.

7.1.2   Wastewater Mist Concentrates

These little airborne beads can convey pathogens and different contaminants. Blenders and aerators make mist concentrates, which aggravate the surface of wastewater tanks and lakes, or by overhead sprinklers. The separation of these mist concentrates can convey in winds and the survival time of pathogens can variable and will rely upon the site. A hazard administration design and control of where and how any treatment plant or land water system region is to found will be essential.

7.1.3   Fluids

The attributes of treated wastewater to be come back to the earth will rely upon the level of treatment it has gotten.

7.1.4 Solids—Muck and Bio Solids

These can name semi-solids and semi-fluids contingent upon the measure of water left in them. Natural solids from essential and optional treatment forms can alluded to as mucks. Nearby specialists, contribute huge exertion into changing over mucks to bio solids and decreasing the level of water in the prepared solids keeping in mind the end goal to make strides dealing with issues when they can arranged to landfills.

7.2 Sorts of Re-entry Framework

Biological communities are dynamic, complex interfacing networks of human, organic and physical forms. Individuals are reliant on characteristic biological communities for the merchandise, administrations and items they give. Therefore, our long haul prosperity is needy on keeping up solid biological systems well into what's to come. The effect of wastewater re-entry on these frameworks will not simply rely upon the amount and nature of residuals discharged into them. It will likewise rely upon the affectability of the biological systems and the relative significance of the environment's merchandise and ventures.

7.3 Solids Re-entry Innovations

On location frameworks: Septage is the pump-out substance from septic tanks, and is weaken and hostile blend of sewage, filth and somewhat processed natural solids. The best methods for taking care of this material is to transport it to a brought together group wastewater treatment plant, where it is prepared in promotion blend with the crude slimes delivered from essential settlement tanks. Where the group plant is an oxidation lake framework, the septage can be added to the facultative lake, however precisely so as not to over-burden the delta zone of the lake with solids. Table 3 shows that Fluid and strong wastewater residuals re-enter the biological community.

Little people group treatment plants utilizing bio filter or initiated muck frameworks create a scope of oozes from the mix of both essential and auxiliary treatment forms. The level of adjustment of these solids by the anaerobic and high-impact forms in the treatment plant decides the volume of last bio solids to be overseen by transfer or usage onto arrive. The wet bio solids might dried on unique sand beds at the treatment plant some time recently gathered as dried 'cake' for trucking to arrive (or even to a strong waste landfill).

Table 3 Fluid and strong wastewater residuals re-enter the biological community

S. no.	Framework	Residuals oversaw
1	Freshwater biological communities (streams, lakes and wetlands)	Treated wastewater emanating (different levels of treatment)
2	Marine biological systems (estuaries, harbours and sea – beach front and seaward)	Treated wastewater emanating (different levels of treatment)
3	Land environments (rural, plant, ranger service or arranged regions)	some untreated wastewater (more uncommon)
4	Atmosphere	(i) Treated wastewater gushing (different levels of treatment) (ii) Smell (iii) Gasses (roundabout and flaring of landfill gasses) (iv) Wastewater pressurized canned products (a result of treatment forms)
5	Landfills (shut frameworks)	Sludges and bio solids
6	Waste-to-vitality plants	dried sludge/bio solids

7.4 Wastewater Emanating Re-entry Innovations

On location frameworks: for these frameworks, the measurements required can dictate by the wastewater amount and quality, and site conditions. Such frameworks must plan and endorsed by a qualified furthermore, experienced individual. The site range taken up by the introduced framework needs to incorporate the space between each trench or hill or water system line, and a cradle zone around the framework impression. In addition, a hold zone ought to put aside adjacent for augmentations to the framework if necessary to deal with startling Framework poor execution because of framework over-burden or abuse.

7.5 Re-utilization of Water and Bio Solids Recovered from Wastewater

Customarily wastewater has overseen as an item that is a danger to both human and biological community wellbeing. Thus, the foundation plan for dealing with such a material will mirror this. Local wastewater contains basic assets, for example, water, supplements and natural material. Treated wastewater produces fluid wastewater and essential and auxiliary ooze, which is the material that remaining parts once the first

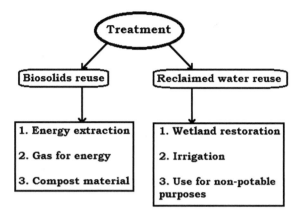

Fig. 6 Alternative reuse strategies

water-borne squander is 'dewatered'. Both these misuses can be set up to recover reusable water and treated the dirt bio solids for plant application as a soil conditioner.

Re-utilization of bio solids requires a larger amount of treatment past what is accomplished with the typical treatment of essential and optional mucks. Various advances can ordinarily utilized that use the asset estimation of wastewater, most ordinarily with brought together frameworks, where the volumes of treated squanders are probably going to be sufficiently huge to empower speculation. It is additionally conceivable with the littler bunch frameworks, despite the fact that this is a genuinely new zone. Re-utilizes incorporate biogas creation for vitality (a procedure that changes over the natural segment of essential and auxiliary slops to methane), water system of water and wastewater supplements for biomass creation, and the utilization of the treated wastewater for wetland rebuilding. Different practices abroad incorporate aquaculture, vitality extraction (from the wastewater) by warm pumps, pee detachment, and supplement stripping for the creation of supplements. The alternative reuse strategies were shown in Fig. 6.

Wellbeing experts likewise have concerns with respect to the utilization of recycled water sourced from wastewater because of the plausibility of direct contact with pathogens if something turns out badly with the treatment procedure, or if the framework is not sufficiently kept up. An extensive variety of advances can investigate, regardless of the possibility that the region is generally new. Like overseeing water use at source, bio solids and recycled water, re-utilize can possibly diminish the general cost of the wastewater framework. For a littler group it might be worth taking a gander at how the waste streams, particularly oozes to change over to bio solids, and may joined with different groups in a brought together process. Re-utilize is well worth investigating as a major aspect of your wastewater considering.

8 Conclusion

The practicality of mechanical, lagoon, and land treatment advancements for wastewater treatment was evaluated, using a course of action of viability pointers made particularly for this examination. The results showed the by and large sensibility of a wastewater treatment advancement is a limit of monetary, environmental and social estimations, and the assurance and comprehension of pointers is affected by a domain's geographic and measurement condition. The delayed consequences of this examination are an undertaking to look past the building expense and characteristic execution related with a particular treatment development in mastermind that decision of an advancement related with the organization of wastewater treatment meets triple standard wants for a comparable modify of natural, money related, and societal sensibility. One target of this paper was to begin a discussion on the most capable strategy to address a more consolidated evaluation of the general reasonability of wastewater treatment developments. It is not as easy to design a wastewater treatment system that regards direction of the workforce, open space, and work in the gathering, besides, limits classy nuances related with smelly air releases, while also constraining costs, imperativeness use, and growing treatment execution.

In the event that a specific wastewater administration technique is considered non-economical, the effect will reach out past its quick operational region and even into future ages. Thus, ordinary practicality markers for wastewater structures that have focused on characteristic stressors at the dismissal of societal issues need to attempt later on to join momentum and intergenerational balanced impacts. In addition, the arrangement of wastewater organization structures that are better planned into greater gathering needs could be considered. For example, the reuse of treated wastewater and organization of solid residuals could be better planned with neighborhood agriculture practices which would re-spread and return supplements back to the incorporating condition, instead of amassing supplement advances in one tolerating water body. In a perfect world the utilization of on location treatment frameworks like septic tanks, built wetlands, and notwithstanding treating the soil restrooms has potential in adding to maintainability as they depend on non-vitality and concoction escalated forms that arrival supplements to the encompassing condition.

References

1. Henriques, J., & Catarino, J. (2017). Sustainable value: An energy efficiency indicator in wastewater treatment plants. *Journal of Cleaner Production, 142,* 323–330.
2. Science Applications International Corporation (SAIC). (2006). *Water and wastewater industry, energy best practice, guidebook, focus on energy.*
3. Hwang, Y., & Hanaki, K. (2000). The generation of CO_2 in sewage sludge treatment systems: Life cycle assessment. *Water Science and Technology, 41*(8), 107–113.
4. Tchobanoglous, G., Stensel, H.D., Tsuchihashi, R., & Burton, F. (2013). *Wastewater engineering: Treatment and resource recovery.* USA: Metcalf and Eddy, Inc., McGraw-Hill. ISBN-13: 978-0073401188.

5. Kumar, P. S., Saravanan, A. (2017). Sustainable waste water treatment methodologies. In *Detox Fashion* (pp. 1–25).
6. Dixon, A., Simon, M., & Burkitt, T. (2003). Assessing the environmental impact of two options for small-scale wastewater treatment: Comparing a reedbed and an aerated biological filter using a life cycle approach. *Ecological Engineering, 20,* 297–308.
7. Kumar, P. S., Saravanan, A. (2017). Sustainable wastewater treatments in textile sector. In *Sustainable fibres and textiles* (pp. 323–346).
8. Carolin, C. F., Kumar, P. S., Saravanan, A., Joshiba, G. S., & Naushad, Mu. (2017). Efficient techniques for the removal of toxic heavy metals from aquatic environment: A review. *Journal of Environmental Chemical Engineering, 5,* 2782–2799.
9. United Nations Commission on Sustainable Development, UNCSD. (1996). *Indicators of sustainable development framework and methodologies*. New York: United Nations.
10. Tsagarakis, K. P., Mara, D. D., & Angelakis, A. N. (2002). Application of cost criteria for selection of municipal wastewater treatment systems. *Water, Air, and Soil pollution, 142,* 187–210.
11. Saravanan, A., Kumar, P.S., Yashwanthraj, M., Sequestration of toxic Cr(VI) ions from industrial wastewater using waste biomass: A review. *Desalination and Water Treatment, 68,* 245–266.
12. Ren, J., & Liang, H. (2017). Multi-criteria group decision-making based sustainability measurement of wastewater treatment processes. *Environmental Impact Assessment Review, 65,* 91–99.
13. Karrman, E. (2001). Strategies towards sustainable wastewater management. *Urban Water, 3,* 63–72.
14. Ding, G. K. C., Ghosh, S. (2017). Sustainable water management—A strategy for maintaining future water resources. *Encyclopedia of Sustainable Technologies,* 91–103.
15. Eriksson, E., Auffarth, K., Henze, M., et al. (2002). Characteristics of grey water. *Urban water, 4,* 85–104.
16. Muttamara, S. (1996). Wastewater characteristics. *Resources, Conservation and Recycling, 16,* 145–159.
17. Deeb, A. A., Stephan, S., Schmitz, O. J., et al. (2017). Suspect screening of micropollutants and their transformation products in advanced wastewater treatment. *Science of the Total Environment, 601–602,* 1247–1253.
18. Annen, G. W. (1972). Efficiency of a grit chamber. *Water Research, 6,* 393–394.
19. O'Melia, C. R. (1997). Coagulation and sedimentation in lakes, reservoirs and water treatment plants. *Water Science and Technology, 37*(2), 129–135.
20. Alfonsin, C., Lebrero, R., Estrada, J. M., et al. (2015). Selection of odour removal technologies in wastewater treatment plants: A guideline based on life cycle assessment. *Journal of Environmental Management, 149,* 77–84.
21. Sillanpaa, M., Ncibi, M. C., Matilainen, A., et al. (2017). Removal of natural organic matter in drinking water treatment by coagulation: A comprehensive review. *Chemosphere, 190,* 54–71.
22. Amini, A., Kim, Y., Zhang, J., et al. (2015). Environmental and economic sustainability of ion exchange drinking water treatment for organics removal. *Journal of Cleaner Production, 104,* 413–421.
23. Pal, P. (2017). Chapter 3—Biological treatment technology. In *Industrial Water Treatment Process Technology* (pp. 65–144).
24. Zhang, Q., Hu, J., Lee, D. J., et al. (2017). Sludge treatment: Current research trends. *Bioresource Technology, 243,* 1159–1172.
25. Kornaros, M., & Lyberatos, G. (2006). Biological treatment of wastewaters from a dye manufacturing company using a trickling filter. *Journal of Hazardous Materials, 136*(1), 95–102.
26. Zhang, Q., Hu, J., & Lee, D. J. (2016). Aerobic granular processes: Current research trends. *Bioresource Technology, 210,* 74–80.
27. Shi, X., Leong, K. Y., & Ng, H. Y. (2017). Anaerobic treatment of pharmaceutical wastewater: A critical review. *Bioresource Technology, 245,* 1238–1244.

Printed in the United States
By Bookmasters